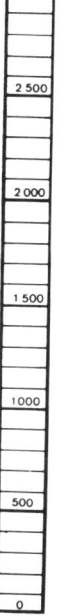

Altitude up to
which the
species occurs

Column

1: period of activity
in north-west
Europe (the rest
of the time the
species
hibernates)

2: breeding period
in north-west
Europe

3: period of activity
in southern
Europe (the rest
of the time the
species
hibernates or
aestivates)

4: breeding period
in southern
Europe

Range

 Species that can be found far
from water

D1827519

Habitat

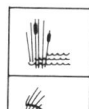 Reed beds in still or running
fresh water

 Beaches, dunes, sandy soils

 Near human habitations and
urbanised areas

 Ruins, walls, verges,
churchyards, open quarries,
sandpits (mostly with a pool)

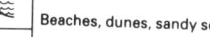 Peat, marshlands, damp
places in moorland

 Deciduous forests

 Pine forests

 Hills, mountains, cliffs

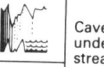 Caverns, karst areas, caves,
underground pools and
streams

 Pasture and arable land,
hedgerows, plains, valleys

 Damp forests in valleys and
plains, alder-brakes

 Meanderings of larger rivers,
lakes and large ponds

 Small water courses, brooks,
running water

 Small water surfaces,
ditches, ponds, marshes,
temporary pools, cattle
watering places, still water

 Artificially created waters

Amphibians of Europe
a colour field guide

D. Ballasina

DAVID & CHARLES
Newton Abbot   London

PHOTOGRAPH CREDITS

J. Van Ammel: 2
S. Frisenda: 1b, c; 4a, b; 9a, b, c; 13b, 14b, d; 18a, b, c; 19; 22a, b; 32a, b; 39; 41a. b; 47d
J. W. Arntzen: 5; 15
I. Baran: 3
H. Strijbosch: 7; 8; 20; 24; 25a, b, c; 29a, c; 30a, b; 35; 42; 45b
J. Wood: 48
D. Ballasina: 1a; 10; 12a, b, c; 13a, c; 14a; 17a, b, c; 22c; 26b, c; 31a, b, c; 32c; 33b; 34a, b, c; 36a, b, c; 38a, b, c; 47a, b, c, e
F. Hordies: 11; 37c; 45a, c
N. Van Esterik: 16b, c
H. Mooijenkind: 16a
P. De Fonseca: 14c; 26a; 29b; 33a, c; 40
K. Grossenbacher: 21; 23; 37; 30c; 41c; 43
Hasler: 28
V. Dupont: 37a, b
H. Wijnands: 44; 46
V. Vandepitte: the picture of the author

Cover illustration: Common Tree Frog *Hyla arborea (D. Ballasina)*
Drawings by Nico Willemsens
Translation by Jan Heyraert

British Library Cataloguing in Publication Data

Ballasina, D.
  Amphibians of Europe.
  1. Amphibians—Europe—Identification
  I. Title   II. Amphibiens. *English*
  597.6'094   QL658.A1

  ISBN 0-7153-8603-4

© De Nederlandsche Boekhandel 1984
© David & Charles (Publishers) Ltd 1984

Filmset by ABM Typographics Limited, Hull
Printed in Italy

CONTENTS

FOREWORD

In recent years much has been published in several European countries on the alarming decline of amphibians. This goes to show that the disappearance of frogs, toads, salamanders and newts is not just an isolated phenomenon but a problem throughout the whole of Europe. At the same time these publications have included demands for protection of these little-known vertebrates.

In spite of many surveys and much new research into their distribution and biology, our knowledge of European amphibians is still far from complete. That is why it is essential to keep recording everyday data on the life of these animals and to pass these records on to interested institutions and conservation societies — for example, by mapping even the smallest breeding ponds or the places where amphibians are found in mountain areas. Only with the help of reliable biological documentation will it be possible to convince the authorities of the importance of preserving small ponds. In the meantime we should take immediate measures to prevent marshes and pools from being drained or filled in.

Salamanders, newts, frogs and toads, almost without exception, play an important part in the food supply of other animals. Moreover, in many ecosystems their numbers are of great importance. Amphibians suffer no less than fish from the increasingly harmful affects of pesticides, with both animal groups being equally dependent on pure water. Their quick response to environmental changes makes most amphibian species important environmental indicators.

The decline of amphibian fauna is an aspect of large-scale changes in the landscape. The destruction of marshes generally results in the extinction of an amphibian population. Protection of species alone (by prohibiting their capture or slaughter) is not effective unless their habitats are also protected. For amphibians this means not only the preservation of their breeding ponds but also their winter and summer habitats.

This book is an excellent means of showing the public at large that protection of amphibians really is essential.

René Honegger
Curator of Herpetology
Vice-chairman of the Species Survival Commission,
International Union for Conservation of Nature and Natural Resources

HOW TO USE THIS GUIDE

Physical characteristics

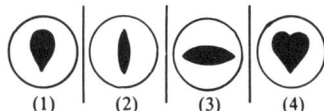

Size of an adult individual (including tail)

Shape of the eye pupil (frogs and toads only)
Triangular (1), Vertical (2), Horizontal (3), Heart-shaped (4)

 Terrestrial species (Toads, Brown Frogs, Tree Frogs): eyes directed sideways

 Aquatic species (Green Frogs and Fire-bellied Toads): eyes directed upwards

 Underside with virtually no markings (Midwife Toads, Alpine Newt, Common Frog and others)

 Underside with markings (Fire-bellied Toads, some Brown Frogs). If the throat alone is marked, symbol is half-coloured

 Poison glands grouped into paratoid glands (*Salamandra, Bufo*): their position is specific for each *Bufo* species

 Two vocal sacs on either side of the head (Green Frogs)

 One vocal sac under the head (Tree Frog, Green Toad, Natterjack)

 Round tails (salamanders)

Tails flattened sideways (newts)

Female lays fertilised eggs singly (oviparous salamanders and newts)

 Female gives birth to living larvae, whether enveloped by a pellicle or not (ovoviviparous salamanders and newts)

 Female gives birth to fully developed and metamorphosed young (salamanders and newts)

 Male carries eggs on back until they hatch

 Axillary mating position or armpit amplexus (Ranidae, Bufonidae, Hylidae)

 Inguinal mating position or loin amplexus (all other frogs and toads)

Egg Forms

 Relatively large clumps deposited without specific form. Mostly the eggs have black or brown upper sides and are enveloped by a layer of jelly. The size of the egg is 1.5–3mm and that of the capsule 6–12mm across (Ranidae, Typical Frogs) Number of eggs per brood: 1,000–12,000

 Eggs are mostly black and are deposited in gelatinous strings 3,000–20,000 (*Bufo bufo, B. viridis*), mostly in two to four rows. When a piece of this string is stretched out carefully the eggs re-arrange into two rows (*Bufo bufo* and *B. viridis*) or one row (*B. calamita*). The eggs of the latter species have paler undersides

 Thick strings with more than one row, wound around plant stems. The eggs are brown or grey (Pelobatidae, Spadefoots). Strings sometimes break into pieces

 Detached eggs deposited singly, sometimes forming a single layer on the pond bed, occasionally on vegetation (*Discoglossus*, Painted Frogs)

 Small clutches (up to 800 eggs), pale yellow to brownish. Size of the egg 1.5mm, that of the capsule 3.4mm. The whole clutch is as big as a walnut

 Eggs of about 2mm, with capsules up to 8mm, deposited singly or in clutches of 15 eggs at most, on the pool bed or on vegetation (Fire-bellied Toads)

Newt eggs, mostly light-coloured, white-yellow, greenish or grey, separate or in very small clutches, wrapped up in the leaf of a water plant. Diameter up to 3mm, in Warty Newts up to 4.5mm. Capsule in some species up to 7mm

Hydromantes, Proteus, Salamandrina, Euproctus, Chioglossa lay their eggs in running water, *Hydromantes* and *Proteus* in caves. *Pleurodeles* deposits its eggs in still water in clutches of 10–15 eggs attached to plants.

Eggs normally on the back of a male. Eggs without visible capsule, pale, joined into strings by narrow links

Habits and Behaviour

Vocal species (most during the spring)

Species that can be found far from water

Solitary (except in mating season)

Gregarious, often with a hierarchical structure

Can be found by day, in full sun

Can be found only in cloudy or damp weather

Nocturnal species

Nocturnal or in heavy rainfall by day

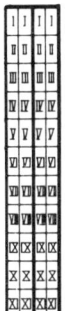

Column 1: period of activity in north-west Europe (the rest of the time the species hibernates)
Column 2: breeding period in north-west Europe
Column 3: period of activity in southern Europe (the rest of the time the species hibernates or aestivates)
Column 4: breeding period in southern Europe

Altitude up to which the species occurs

European range of the species

Habitat

 Reed beds in still or running fresh water

 Beaches, dunes, sandy soils

 Near human habitations and urbanised areas

 Ruins, walls, verges, churchyards, open quarries, sandpits (mostly with a pool)

 Peat, marshlands, damp places in moorland

 Deciduous forests

 Pine forests

 Hills, mountains, cliffs

 Caverns, karst areas, caves, underground pools and streams

 Pasture and arable land, hedgerows, plains, valleys

 Damp forests in valleys and plains, alder-brakes

 Meanderings of larger rivers, lakes and large ponds

 Small water courses, brooks, running water

 Small water surfaces, ditches, ponds, marshes, temporary pools, cattle watering places, still water

 Artificially created waters, drainage basins of motorways and factories, flood dam lakes, park ponds, neglected swimming pools, water-troughs, rainwater cisterns

INTRODUCTION

The main object of this book is to meet the increasing interest in this fascinating group of animals. Unfortunately this growing interest is matched by the steady decline of amphibians in their natural environment.

The guide is intended especially for those people who are aware of the value of amphibians and are alarmed at their disastrous decline in the last two decades — novices as well as students and experienced naturalists. With the help of the guide's expert pictograms and photographs, enthusiasts will be able to examine European amphibians in their natural state. The subjects have been photographed as far as possible in the open, even though terrarium pictures might have been technically more perfect. Newts were of course photographed in aquariums.

The pictograms have been designed so that the observer gains an immediate visual guide to the characteristics, habits, range and habitat of the animal concerned. While it is not intended that these pictograms should be a complete substitute for traditional field guides, nevertheless this compact, at-a-glance means of quick identification and information will be found to be most useful.

Most books on herpetology deal with both amphibians and reptiles, but this guide is confined to amphibians in order to present every European species in detail with several photographs and a clear pictogram of each — a major advantage for both novice and expert compared to other guides on this subject.

SALAMANDERS AND NEWTS, FROGS AND TOADS: A WORLD APART

Not so long ago an increasing number of people began to take a positive interest in this animal group. For centuries, however, amphibians had been the object of revulsion and fear and they found their place in legends, horror stories and fairy-tales about princes changing into frogs and witches using toads in their spells.

Most children have searched for tadpoles in the spring in order to follow the wonderful metamorphosis from tadpole to frog in an aquarium or jamjar. Each spring these animals move to open water without which, with a few exceptions, they cannot survive. The word 'amphibian' is derived from this dual way of life; 'amphi' is ancient Greek for 'dual' and 'bios' for 'life'. In general they lead a larval life in water and an adult life on land. For breeding purposes most species require an aquatic environment and it is usually during this breeding stage that the greatest number of amphibians are encountered.

Male newts, often with large and beautiful crests, display before females in their nuptial dress. During this mating period the males of different species of frogs and toads are often heard croaking, grunting, squeaking or whistling in or near the water. Some species have vocal sacs that serve as resonators or amplifiers: either two as in Green Frogs (*Rana esculenta* complex) which produce a croaking sound, or just one vocal sac under the head as in Green Toads (*Bufo viridis*) which produce a mysteriously trilled whistling, in Natterjacks (*Bufo calamita*) which squeak very loudly like a kind of giant swift or in Tree Frogs (*Hyla arborea* and *H. meridionalis*) with their very loud 'keck-keck-keck-keck', often audible up to a distance of about one kilometre.

Other species make a more modest noise: most Brown Frogs produce a soft croak (as does *Rana dalmatina*) or a grunt (*Rana temporaria*) or even the sound of air escaping from a submerged bottle (*Rana arvalis*). Some species emit an almost indescribable sound akin to the tinkling or the soft tolling of bells (Midwife Toads, *Alytes obstetricans* and *A. cisternasii*; the Parsley Frog, *Pelodytes punctatus*; and Fire-bellied Toads, *Bombina bombina* and *B. variegata*). Apart from this the latter inflate their whole bodies as resonators for their faint voices. Some other species (Spade-foots, *Pelobates* sp.) emit a 'c'lock-c'lock' under water.

The animals breed in water, often by laying very large numbers of eggs, ranging from a few hundred per season to over 12,000.

In summer most young toads and frogs leave their ponds and say farewell to their aquatic life. In order not to run the risk of dehydration the newly metamorphosed amphibians mainly leave the water during warm summer rains, and at times incredible numbers of small frogs and toads spread all over the surrounding countryside. Man has always been startled by this phenomenon without being able to find an explanation for it; thus these little creatures 'fell from heaven', or it 'rained frogs and toads'. Hence in part the origin of the belief in frog rains and other mysterious stories linked to the world of toads, frogs, salamanders and newts.

The venomousness of some species, too, has been known to man for a long time, although it has often been greatly exaggerated. A typical example is the brightly coloured black-and-yellow Fire Salamander (*Salamandra salamandra*). Because of the magic qualities ascribed to amphibians, these animals have been used for centuries in brewing medicines and even aphrodisiacs.

THE ORIGIN AND EVOLUTION OF AMPHIBIANS

Although it is not the aim of this book to give a complete palaeontological survey of the origin and evolution of amphibians, the importance of these animals as a link in the evolution towards higher vertebrates including man should be emphasised. If amphibians had not taken the plunge towards a more land-based way of life, the evolution of higher vertebrates, as we know it, would not have been possible. What made the amphibians change their habitat is still not clear. In both the animal and vegetable kingdoms, we meet examples of an often unknown force urging individuals to leave their normal environment.

With amphibians this change is quite dramatic: from aquatic life they adapt to terrestrial life, a shift which repeats itself every spring with great rapidity — within a few weeks or months — whereas it took their ancestors millions of years to accomplish it. Yet amphibians as a group have never achieved complete independence of water: they still need it (or at least very damp places) to breed. Reptiles, which originate from amphibians, were the first vertebrates able to become fully independent of an aquatic environment.

It is thought that amphibians developed out of a specific group of fish (*Crossopterigii*) during the middle of the Devonian period 370 million years ago. These fish had paired fins which developed in amphibians into two pairs of limbs with separate fingers and toes. The great expansion of the amphibians took place no earlier than the Carboniferous age. They were the first animals to conquer the land and thus suffered little competition from other animal groups, which enabled them to develop in all possible directions. At the end of this period 300 million years ago the reptiles, from which birds and mammals later originated, came into being and ushered in the decline of the amphibian period at the end of the Permian age.

Nowadays we know of just under 3,000 living amphibian species — about 300 salamandrids (Caudata or Urodela: tailed amphibians) and about 2,700 toads and frogs (Anura or Salientia: tailless amphibians). No more than 50 species, all of which are in this guide, are to be found in Europe, from the British Isles to the Urals. These represent but a tiny fraction of the past, rich in species and forms, but perhaps this is what makes them so interesting.

THE ANATOMY AND CLASSIFICATION OF AMPHIBIANS
Anatomy

Although the evolution from aquatic to terrestrial life repeats itself each year with the metamorphosis of amphibian larvae, these tadpoles do not resemble their fish ancestors at all. For example, tadpoles do not have fins, and the metamorphosis into a terrestrial animal substantially differs in its mechanism from the fish-to-amphibian evolutionary line. The differences between present-day amphibians and fish represent more than just the difference between gills and fins in fish on the one hand and lungs and legs in amphibians on the other.

Strictly speaking there is no characteristic typical of amphibians alone, as are feathers for birds or mamilla for mammals. Therefore it is important to gain an insight into the anatomical structure of amphibians and to summarise the differences compared to other vertebrates; this is also the best way to understand their physical characteristics.

Skin

The change from aquatic to terrestrial life had enormous anatomical consequences for amphibians. In the first place the smooth, naked amphibian skin replaced the scaly skin of fish. The amphibian skin, as in other vertebrates, consists of two layers: the epidermis and the dermis. The epidermis consists of several layers of cells (two at least in

11

amphibian larvae that have not yet left the egg). The outermost layer (the stratum cornicum) is horny and contains dead cells which help to avoid dehydration. This is a typical characteristic of land animals and is not found in a number of salamander species, for example Olms (*Proteus anguinus*) which spend their entire life under water. This regular shedding and replacing of skin is called 'sloughing'. In newts especially the process is easy to follow and a virtually complete transparent 'empty' newt can be seen floating in the water. In frogs and toads the process is the same, but it is almost impossible to find a toad skin: in sloughing the toad makes a kind of bending movement so that the outermost skin layer is torn open; it is then stripped off as a whole with the help of fore and hind limbs, after which it is eaten.

The dermis is situated under the epidermis and is much thicker. It consists of two layers: the innermost, rather compact layer (stratum compactum) and the outermost layer (stratum spongiosum) which is very rich in mucous and venomous glands and in pigment cells (chromatophora) which are responsible for the body colours and colour changes by means of brown, yellow and red dyestuffs and guanine cells. A large number of blood vessels is present which is only logical since the skin, together with the lungs, also serves as a respiratory organ. The glands situated in the skin start to function in the early larval stage. For example, glands at the head of the larva secrete enzymes to produce a local loosening of the egg capsule, through which the larva can leave the egg. Unlike glands in fish, amphibian glands are mostly multicellular. In certain species (salamanders, *Salamandra* sp., and Typical Toads, *Bufo* sp.) multicellular venomous glands are well developed and grouped into paratoid glands (see the pictograms for the species concerned).

As distinct from fish, reptiles or birds, hornified skin virtually never occurs in amphibians. In Spadefoots (*Pelobates* sp.), however, there is a hoe-like horny tubercle on the hind foot which is used in digging. This is the only exception in European species.

Skeleton

Amphibians have the skeleton of a typical land animal. Since adult amphibians are no longer supported by water, as fish still are, they had to develop a bony skeleton (endoskeleton or internal skeleton). In their larvae it still consists mainly of cartilage.

Within the amphibian skeleton we can distinguish the skull, spine and ribs, breastbone (sternum), shoulder- and hip-girdles and limbs. Depending on the way of life of the different species there are quite a number of differences in skull form. Generally speaking the amphibian skull, quite unlike the skull of a fish, is horizontally flattened, rather broad and with virtually no covering bones (though these often appeared in fossil species); it contains a braincase (chondrocranium); the lower jaw is attached to the skull, which is not the case with fish.

In fish the dorsal vertebrae are very similar from skull to tail and they cannot turn. Amphibians are the first animals in which the vertebrae differ according to function: carrying and moving the head (cervical vertebrae, with the beginning of an atlas vertebra on which the head turns); carrying the trunk (eight thoracic vertebrae in frogs

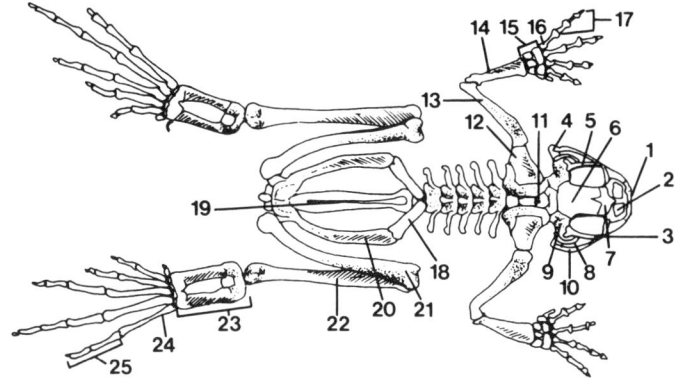

Fig 1 Skeleton of a frog
1 Intermaxillary; 2 Nasal bone; 3 Jaw bone; 4 Quadrate; 5 Pterygoid; 6 Fronto-parietal; 7 Sphenetmoid; 8 Squamosal; 9 Prootic bone and exoccipital; 10 Jugal; 11 Atlas; 12 Shoulder blade; 13 Humerus; 14 Radius and ulna; 15 Carpal bones; 16 Metacarpal bones; 17 Finger phalanges; 18 Sacral vertebra; 19 Urostyle; 20 Ilium; 21 Femur; 22 Splint-shin-bone; 23 Tarsal bones; 24 Metatarsal bones; 25 Toe phalanges

and toads); adaptations for shoulder- and hip-girdles (including the sacral vertebra to which the strong hind legs are attached in frogs); and finally caudal vertebrae (missing in frogs and toads).

The ribs are poorly developed in present-day amphibians: in salamanders and newts they are very short and in frogs and toads they may even be absent or else they are virtually unossified (cartilage only). They never constitute a closed chest and are not attached to the breastbone (the sternum) which among vertebrates appears for the first time in amphibians. Newts and salamanders only develop a rather primitive cartilaginous sternum, but in frogs and toads it is quite well-developed and at the same time partly ossified.

Shoulder- and hip-girdles are also found for the first time in amphibians. This is the major feature distinguishing the superclass Tetrapoda (four limbs), which is subdivided into the classes of Amphibia, Reptilia, Aves and Mammalia. European amphibians have four fingers on the fore legs and five toes on the hind legs.

Digestive Organs

The amphibian digestive system is only slightly different to that of fish, but the tongue has undergone a dramatic change. This development appears to be characteristic of the shift to terrestrial life, in which prey are not necessarily already wet as they are under water. Thus land animals have to moisten their food by turning it around in the mouth cavity and chewing it. The tongue is rich in glands: in Typical Frogs and Typical Toads (*Rana* and *Bufo*) it is attached to the front of the mouth cavity and can be flicked forwards. At the same time it is very sticky, having developed into a catching device. The prey (insect, snail or worm) is caught by the flicking tongue and pulled into the mouth.

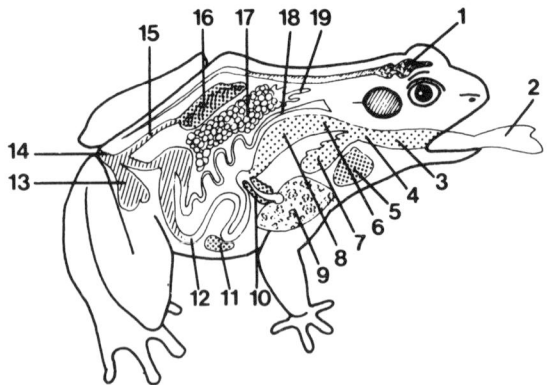

Fig 2 Entrails of a frog
1 Brain; 2 Tongue; 3 Ear-drum; 4 Larynx; 5 Heart; 6 Esophagus; 7 Lung;
8 Stomach; 9 Liver; 10 Pancreas; 11 Spleen; 12 Intestine; 13 Urinary bladder;
14 Cloaca; 15 Urethra; 16 Kidney; 17 Ovary; 18 Oviduct; 19 Yellow corpuscle

In other toads (Discoglossidae such as *Alytes* sp., *Bombina* sp., *Discoglossus* sp.) the tongue is disc-shaped and cannot be flicked forwards. This is one of the characteristics on which the classification of frogs and toads has been based. In other more primitive species (Archaeobatrachia) the disc-shaped tongue can be flicked forwards (the separation of this group is based on, among other things, the absence of ribs); examples are the family of Spadefoots (Pelobatidae: *Pelobates* sp. and *Pelodytes* sp.).

Amphibians have teeth too (with the exception of Typical Toads, Bufonidae, in Europe); usually they are situated on the upper jaws and palate. The rest of the digestive tract ends in a cloaca, a cavity which constitutes the common end of the digestive, excretive and reproductive systems.

Respiratory System

During initial stages amphibian larvae breathe through external gills, soon replaced by internal gills in frog and toad larvae. Among salamanders and newts the gills are preserved only in Olms (*Proteus anguinus*) or in the case of neoteny, when the animal keeps its larval appearance, but is able to breed. In all other cases the European species grow lungs and resorb the gills during metamorphosis.

Since the ribs in amphibians are developed poorly or not at all, and since they lack a diaphragm, these animals have developed a special respiratory technique; in effect they 'swallow' air, which is pressed into the lungs by movements of the tongue and throat. The more fresh oxygen is needed, the faster this happens.

Within the tract to the lungs is the larynx in which the vocal cords are situated: this development makes amphibians the first higher animals to emit sounds. In frogs and toads the sound is amplified — but not produced, as people often mistakenly assume — by external vocal sacs (as in the *Rana esculenta* complex, *Bufo viridis*, *Hyla* sp.) or, in a few instances, by expansion of the internal vocal sacs (eg in *Bufo bufo*).

The lungs then are relatively simple structures compared to higher vertebrates. Oxygen requirements are met to a great extent by skin respiration. When hibernating underwater or, in some species, during the mating season, the animals depend totally on skin respiration.

Circulatory system

Amphibians are cold-blooded (poikilothermic) animals, which means they are entirely dependent on the temperature of their environment. Instead of a simple heart with two auricle cavities as in fish, the amphibian heart has three cavities: two heart auricles and one heart ventricle. The blood that has flowed through the entire body goes back to the heart and enters the right auricle, whereas the left auricle receives the blood that comes back from the lungs. As a consequence there is a moderate fusion of reoxygenated and deoxygenated blood (unlike the fourfold heart of birds and mammals, which have separate auricles and ventricles). In amphibians this fusion is limited to a great extent by a system of valves, better developed in toads and frogs than in salamanders and newts. During metamorphosis the circulatory system changes dramatically, since in larvae the oxygen is obtained through gills and in adult animals through lungs.

Reproductive system

The reproductive system of frogs and toads is illustrated in Fig 2. In salamanders and newts as well as the cloacal cavity there is a second cavity for sperm storage (spermatheca). Fertilisation in salamanders and newts is in fact internal; in frogs and toads it is external, as explained in the section on behaviour and in the pictograms. Stages in larval development are shown in Figs 3a and 3b.

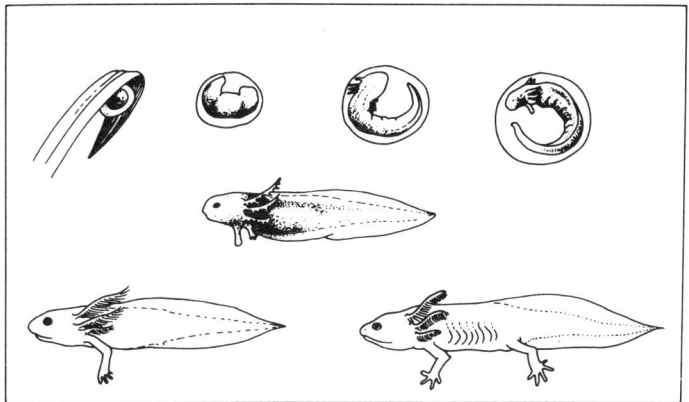

Fig 3a Larval development of salamanders and newts

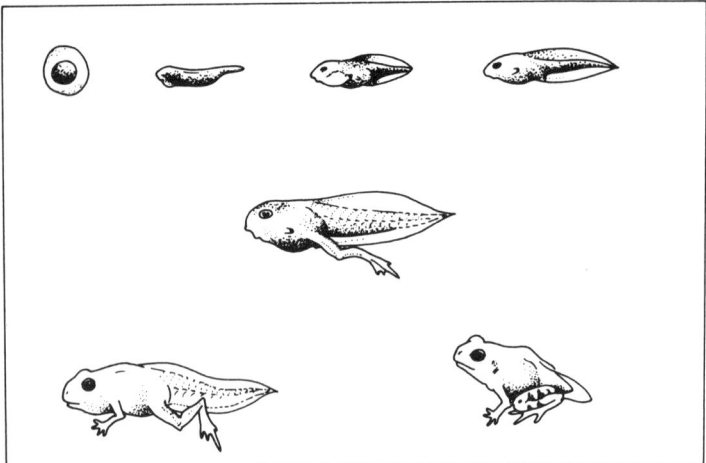

Fig 3b Larval development of frogs and toads

Sensory organs

The senses changed dramatically while adapting to terrestrial life. The special sense-organ of fish for perceiving vibrations in the water survives only in amphibian larvae and in fully aquatic species such as the Olm (*Proteus anguinus*).

The eyes are improved and are provided with lachrymal glands and closeable eyelids. These are absent in larvae. In many frogs and toads the protruding eyeballs are involved in helping to swallow prey, since they are separated from the mouth cavity by only a few layers of tissue.

The ears, too, are quite well developed in frogs and toads. In frogs a large eardrum is often very clearly visible. Soil vibrations are transmitted to the ear though the fore limbs and shoulder-girdle.

Finally, amphibians were the first animals to develop Jacobson's organ, which is highly developed in reptiles. It is part of the olfactory system, which is very important in the orientation of the animal and in the localisation of prey.

Classification of Amphibians

Living amphibians are divided into three orders: Urodela (salamanders and newts); Apoda (caecilians — limbless amphibians, not found in Europe); and Anura (toads and frogs).

European amphibians can be classified as follows (groups without European representatives have been omitted):

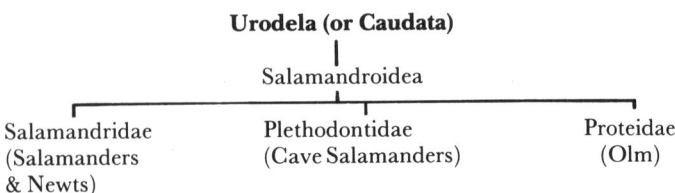

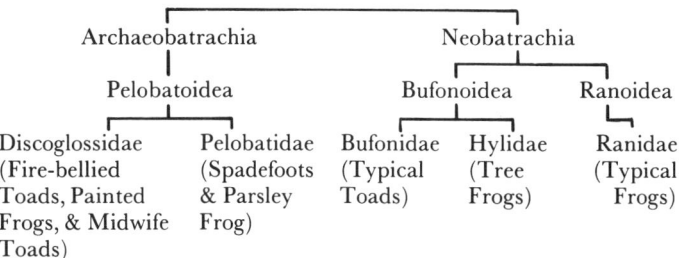

Anura (or Salientia)

Archaeobatrachia		Neobatrachia		
Pelobatoidea		Bufonoidea		Ranoidea
Discoglossidae (Fire-bellied Toads, Painted Frogs, & Midwife Toads)	Pelobatidae (Spadefoots & Parsley Frog)	Bufonidae (Typical Toads)	Hylidae (Tree Frogs)	Ranidae (Typical Frogs)

To show how the system works, here is the full classification of the Common Tree Frog (*Hyla arborea arborea*):

PHYLUM:	Chordata
SUBPHYLUM:	Vertebrata
SUPERCLASS:	Tetrapoda (quadrupeds)
CLASS:	Amphibia
ORDER:	Anura (tailless amphibians)
SUBORDER:	Neobatrachia (later amphibians)
SUPERFAMILY:	Bufonoidea (toad-like)
FAMILY:	Hylidae (Tree Frogs)
GENUS:	*Hyla*
SPECIES:	*arborea*
SUBSPECIES:	*arborea*

One thing this example shows is that there is no scientific difference between a 'frog' and a 'toad'. In general most people call *Bufo, Alytes* and *Pelobates* species 'toads', whereas *Rana* and *Hyla* species are called 'frogs'. *Pelodytes* species are sometimes called 'toads', sometimes 'frogs', but such popular classifications have no zoological basis at all.

Pages 18–19 give a list by family of every species covered in this guide. For readers who travel abroad or wish to consult foreign books on the subject, it may be useful to know European vernacular names too.

DISTRIBUTION AND ENVIRONMENT
Distribution

With 3,000 species in existence today, amphibians are still a successful group of vertebrates. Whilst salamanders and newts occur predominantly in the temperate regions of North America and Europe, ie within a relatively restricted range, toads and frogs are found in all continents except Antarctica. They are mainly found in damp areas such as tropical forests where they take on an incredible variety of forms. However, frogs and toads are also found in unlikely areas for amphibians such as deserts and high mountains, where they need to adapt in special ways in order to survive successfully.

In Europe there is a substantial difference between the number of species in the north-west and in the south. Southern Europe has far more amphibian species, and the populations of those species that also

LATIN	ENGLISH	FRENCH	GERMAN	ITALIAN	SPANISH	DUTCH
Salamandridae	**Salamanders and newts**	**Salamandres and tritons**	**Salamanderen and Wassermolchen**	**Salamandre and tritoni**	**Salamandras and tritones**	**Salamanders**
Salamandra salamandra	Fire Salamander	Salamandre, Salamandre terrestre, Salamandre tachetée, Salamandre commune	Feuersalamander	Salamandra, Salamandra pezzata	Salamandra, Salamandra común	Vuursalamander, Landsalamander
Salamandra atra	Alpine Salamander	Salamandre noire	Alpensalamander	Salamandra alpina, Salamandra nera	Salamandra alpina	Alpen (land)salamander
Salamandra (Mertensiella) luschani	Luschan's Salamander	Salamandre de Luschan	Lycischer Salamander	Salamandra di Luschan	Salamandra de Licia	Egeïsche landsalamander
Salamandrina terdigitata	Spectacled Salamander	Salamandrine à trois orteils	Brillensalamander	Salamandrina dagli occhiali	Salamandra de anteojos	Brilsalamander
Chioglossa lusitanica	Golden-striped Salamander	Salamandre portugaise or Salamandre à bandes dorées	Goldstreifensalamander	Salamandra a strisce dorate	Salamandra rabilarga	Goudstreepsalamander
Pleurodeles waltl	Sharp-ribbed Salamander	Pleurodèle de Waltl	Spanischer Rippenmolch	Pleurodele or tritone di Waltl	Gallipato	Ribbensalamander
Euproctus asper	Pyrenean Brook Salamander	Triton des Pyrénées	Pyrenäen-Gebirgsmolch	Tritone or euproto dei Pirenei	Tritón pirenaico	Pyreneeënbeeksalamander
Euproctus platycephalus	Sardinian Brook Salamander	Triton de montagne de Sardaigne	Sardinischer Gebirgsmolch	Tritone or euproto sardo	Tritón de Cerdeña	Sardijnse beeksalamander
Euproctus montanus	Corsican Brook Salamander	Triton de montagne de Corse	Korsischer Gebirgsmolch	Tritone or euproto corso	Tritón de Córsega	Corsicaanse beeksalamander
Triturus marmoratus	Marbled Newt	Triton marbré	Marmormolch	Tritone marmorato	Tritón jaspeado	Marmersalamander
Triturus cristatus	Warty Newt	Triton crêté	Kammolch	Tritone crestato	Tritón crestado	Kamsalamander
Triturus carnifex	Italian Warty Newt	Triton crêté italien	Italienischer Kammolch	Tritone crestato italiano	Tritón crestado italiano	Italiaanse kamsalamander
Triturus alpestris	Alpine Newt	Triton alpestre	Bergmolch	Tritone alpestre	Tritón alpino	Alpen (water)salamander
Triturus boscai	Bosca's Newt	Triton de bosca	Spanischer Wassermolch	Tritone di Bosca	Tritón ibérico	Spaanse watersalamander
Triturus helveticus	Palmate Newt	Triton helvétique or palmé	Fadenmolch	Tritone palmato	Tritón palmeado	Vinpootsalamander
Triturus vulgaris	Smooth Newt	Triton ponctué, triton vulgaire	Teichmolch	Tritone punteggiato	Tritón punteado	Kleine watersalamander
Triturus italicus	Italian Newt	Triton italien	Italienischer Wassermolch	Tritone italiano	Tritón italiano	Italiaanse watersalamander
Triturus montandoni	Montandon's Newt	Triton de Montandon	Karpatenmolch	Tritone di Montandon	Tritón de los Cárpatos	Karpatensalamander
Plethodontidae	**Cave Salamanders**	**Salamandres cavernicoles**	**Höhlensalamanderen or Schleuderzungen-Salamanderen**	**Geotritoni**	**Salamandras cavernícolas**	**Grottensalamanders**
Hydromantes genei	Sardinian Cave Salamander	Salamandre cavernicole de Sardaigne	Höhlensalamander or Sardinischer Schleuderzungen-Salamander	Geotritone sardo	Salamandra cavernícola de Cerdeña	Sardijnse grottensalamander
Hydromantes italicus	Italian Cave Salamander	Salamandre cavernicole d'Italie	Italienischer Höhlensalamander or Italienischer Schleuderzungen-Salamander	Géotritone italiano	Salamandra cavernícola italiana	Italiaanse grottensalamander
Proteidae	**Olms**	**Protées**	**Olmen**	**Protei**	**Proteos**	**Olmen**
Proteus anguinus	Olm	Protée	Olm	Proteo	Proteo	Olm

Scientific name	English	French	German	Italian	Spanish	Dutch
Discoglossidae	**Painted Frogs, Midwife Toads and Fire-bellied Toads**	**Discoglossidés: les crapauds discoglosses, les crapauds sonneurs and les crapauds accoucheurs or alytes**	**Scheibenzüngler**	**Discoglossidi**	**Sapos parteros, sapos de vientre de fuego and sapillos pintojos**	**Echte Schijftongkikkers, vroedmeesterpadden and vuurbuikpadden**
Discoglossus pictus	Painted Frog	Discoglosse peint	Gemalter Scheibenzüngler	Discoglosso dipinto	Sapillo pintojo	Schijftongkikker
Discoglossus sardus	Tyrrhenian Painted Frog		Sardinischer Scheibenzüngler	Discoglosso sardo	Sapillo pintojo de Cerdeña	Sardijnse schijftongkikker
Alytes obstetricans	Midwife Toad	Alyte or crapaud accoucheur	Geburtshelferkröte	Alite ostetrico	Sapo partero commun	Vroedmeesterpad
Alytes cisternasii	Iberian Midwife Toad	Crapaud accoucheur d'Espagne or alyte de Cisternas	Iberische Geburtshelferkröte	Alite iberico	Sapo partero ibérico	Spaanse vroedmeesterpad
Bombina variegata	Yellow-bellied Toad	Sonneur à pieds épais or sonneur à ventre jaune	Gelbbauchunke	Ululone a ventre giallo	Sapo de vientre amarillo	Geelbuikvuurpad
Bombina bombina	Fire-bellied Toad	Sonneur à ventre de feu	Rotbauchunke	Ululone a ventre rosso	Sapo de vientre de fuego	Roodbuikvuurpad
Pelobatidae	**Spadefoots**	**Pélobatidés**	**Krötenfrösche**	**Pelobatidi**	**Sapos de espuelas and sapillo moyeado**	**Knoflookpadden and groengestipte pad**
Pelobates fuscus	Common Spadefoot	Pélobate brun	Knoblauchkröte	Pelobate fosco	Sapo de espuelas pardo	Knoflookpad
Pelobates cultripes	Western Spadefoot	Pélobate or crapaud à couteaux	Messerfusz	—	Sapo de espuelas	Spaanse knoflookpad
Pelobates syriacus	Eastern Spadefoot	—	Syrische Schaufelkröte	—	Sapo de espuelas oriental	Syrische knoflookpad
Pelodytes punctatus	Parsley Frog	Pélodyte ponctué	Westlicher Schlammtaucher	Pelodite punteggiato	Sapillo moteado	Groengestipte pad
Hylidae	**Tree Frogs**	**Grenouilles arboricoles**	**Laubfrösche**	**Ilidi or raganelle**	**Ranitas de San Antonio**	**Boomkikkers**
Hyla arborea	Common Tree Frog	Rainette verte or graisset	Laubfrosch	Raganella commune	Ranita de San Antonio	Europese boomkikker
Hyla meridionalis	Stripeless Tree Frog	Rainette méditerranéenne	Mittelmeer-Laubfrosch	Raganella mediterranea	Ranita meridional	Mediterrane boomkikker or Zuideuropese boomkikker
Bufonidae	**Typical Toads**	**Crapauds typiques**	**Kröten**	**Bufonidi or rospi tipici**	**Sapos típicos**	**Echte padden**
Bufo bufo	Common Toad	Crapaud brun or commun (Crapaud)	Erdkröte	Rospo commune	Sapo commun	Bruine or gewone pad
Bufo calamita	Natterjack	calamite or crapaud des joncs	Kreuzkröte	—	Sapo corredor	Rugstreeppad
Bufo viridis	Green Toad	Crapaud vert	Wechselkröte	Rospo smeraldino, rospo verde	Sapo verde	Groene pad
Ranidae	**Typical Frogs**	**Grenouilles typiques**	**Frösche**	**Ranidi or rane tipiche**	**Ranas típicas**	**Echte kikkers**
Rana temporaria	Common or Grass Frog	Grenouille rousse	Grasfrosch	Rana temporaria or rana alpina	Rana bermeja	Bruine kikker
Rana arvalis	Moor Frog	Grenouille oxyrhine or grenouille agile	Moorfrosch	Rana arvale	Rana campestre	Hei (de) kikker
Rana dalmatina	Agile Frog	Grenouille agile or grenouille rousse	Springfrosch	Rana agile	Rana ágil	Springkikker
Rana latastei	Italian Agile Frog	Grenouille agile d'Italie	Italienischer Frosch	Rana di Lataste	Rana de Lataste	Italiaanse kikker
Rana graeca	Stream Frog	Grenouille grecque	Griechischer Frosch	Rana greca	Rana griega	Griekse kikker
Rana iberica	Iberian Frog	Grenouille ibérique	Seefrosch	Rana iberica	Rana patilarga	Spaanse kikker
Rana ridibunda	Marsh or Lake Frog	Grenouille rieuse		Rana ridibonda or rana verde maggiore	Rana verde commun	Grote groene kikker, Meerkikker
Rana perezi	Southern Marsh Frog	Grenouille rieuse d'Espagne	Spanischer Seefrosch	—	Rana verde commun	Spaanse meerkikker
Rana lessonae	Pool Frog	—	Teichfrosch	Rana (verde) di Lessona	Rana verde europea	Kleine groene kikker
Rana esculenta	Edible Frog	Grenouille verte or grenouille comestible	Teichfrosch	Rana esculenta or rana verde minore	Rana verde comestible	Groene kikker
Rana catesbeiana	American Bullfrog	Grenouille-taureau d'Amérique	Amerikanischer Stierfrosch	Rana toro	Rana toro americana	Stier- or brilkikker

occur in the north are far more numerous, often larger and with a positive population structure (ie relatively more young animals). However, north-west Europe has quite large, though often very local, populations of certain species that are less sensitive to cold. What are the reasons for this phenomenon?

Climate

This is probably the decisive factor. Amphibians being poikilothermic but, compared to fish, relatively thermophile, their distribution is determined mainly by temperature and moisture, which are characteristics of a specific climatological type.

Apart from physically insuperable natural obstacles — such as seas, large rivers, or mountain ranges — too dry, too cold or too changeable a climate can also prevent the distribution of a species. The climate in Europe over the past three million years has been relatively unstable: there have been four ice ages during which ice covered a large part of north-west Europe, making virtually all life impossible.

Amphibians were able to maintain themselves only in regions where suitable climatological conditions occurred: Iberia (Spain and Portugal), the Italian peninsula (Italy and its offshore islands) and the Balkans (Yugoslavia, Albania, Greece, Bulgaria and Rumania). The consequent geographical isolation resulted in new forms, subspecies and often entirely separate species.

With the retreat of the ice cap the climate began to grow warmer, and amphibians were able to expand their distribution northward, some species — such as the Common Frog (*Rana temporaria*) — reaching beyond the Arctic Circle. This expansion gave rise to what we know as the Green Frog complex (qv).

Human factors

Man has exerted an enormous influence on European nature. The landscape of Europe is determined largely by cultivated land. There are only a few dispersed semi-natural areas and genuinely natural areas are extremely rare.

At first man actually increased the number of vegetable and animal species because his activities resulted in a diversification of the landscape and hence a greater variety of habitats within a relatively small area. The turning-point came in medieval times: deforestation took on alarming proportions, marshes were drained everywhere and amphibians were used for food during Lent and famines. They were also hunted because of their evil reputation. In addition the climate has grown cooler since the Middle Ages, so that several species in north-west Europe have only been able to maintain themselves in micro-climatologically favourable enclaves, where the influence of soil and topography has created suitable environments despite the prevailing climate.

Since the eighteenth century the impact of industrialisation on nature has accelerated. But the greatest blow did not come until the present century, with the vast expansion of urbanised areas and their accompanying roads and traffic, particularly after the Second World

War. Other factors include increasing pollution from the chemical industry, the massive use of fertilisers (causing eutrophication of water) especially in western Europe, and herbicides and insecticides which have a disastrous effect on amphibians. All of this has resulted in an appalling decline in each species, especially in areas with the least favourable climatological conditions, which has been clearly notice-able since the end of the sixties. Lowlands have suffered most from these factors (see the section on threats to our herpetofauna, page 28).

In general, amphibian populations are less endangered in southern Europe because the climate is more favourable, the landscape is less accessible to industry and large-scale agriculture, and the human population is less dense than in north-west Europe. However, even in the south a relatively sharp decline has been detected.

Environment

Amphibians can be found in virtually any place containing fresh water for breeding. This includes still water — in watering places for cattle (*Bufo* sp.) or even concrete water troughs (*Hyla* and *Triturus* sp.), flooded cart ruts (*Rana temporaria*), ditches (*Rana esculenta* complex), ponds, pools, marshes, alder-brakes (a number of newts, salamanders, toads and frogs) and lakes (*Triturus cristatus, Rana esculenta, R. ridibunda*) — also slowly-running water (*Rana esculenta, Triturus* sp.) and rapidly-flowing brooks (*Salamandra* sp.). Sometimes a species even adapts to slightly brackish water (Natterjack, *Bufo calamita*).

More detailed information on each species can be found in the pictograms. Adult terrestrial amphibians often remain attached to wet habitats outside the breeding season and they are often nocturnal, come out only in damp rainy weather, or live in the shade under vegetation or rocks, in weeds or even caves. The pictograms will provide more information on all these habitats.

THE BEHAVIOUR OF AMPHIBIANS

The habits of amphibians are far from well-known, particularly those of European species.

Migrations

Surprisingly amphibians migrate, as do some birds and mammals. In general this is not a conspicuous phenomenon because migrations mainly take place in rainy weather and at night. Usually the only evidence that remains is the remnants of animals that have been run over on the road.

Spring migration

These migratory movements are probably the best known. Under the influence of hormonal changes, stimulated by moisture and tempera-ture, the sexually mature animals move to the breeding ponds, a distance of some 50 to 3,500 metres from the neighbouring area in which they live. The time at which they start to move is determined by moisture and temperature in some species, in others by a kind of internal clock. In some species (for example *Bufo bufo*) the temperature for mating activity is around 5 to 6°C, after which the animals burrow and remain inactive in their summer habitats until the temperature rises above 10 to 12°C.

Autumn migration

Autumn migration is a less well-known phenomenon. It is not on such a large scale and hence is less conspicuous, a fact which may cause difficulties when taking protective measures (see the section on protection). In autumn the animals move from their summer habitats to the places where they hibernate, which may be the same as their breeding grounds. Occasionally individuals of *Bufo bufo* and *B. viridis* may mate in the autumn, and in the case of the latter, may actually breed.

Most species have a permanent summer habitat to which they move immediately after mating in order to burrow and await warmer weather. They can be found there from March until September. Young animals occupy different summer territories to older animals, mostly in dense vegetation, thereby largely avoiding cannibalism. In any case the animals seek out a permanent refuge to which they always return. This desire for a safe refuge can be observed in terraria, where too little space can produce stress in the animals, which in turn may result in refusal of food and death.

Besides migrations, which are specific and recurrent, genuine emigrations also exist (such as those of *Hyla arbora, H. meridionalis,* and *Bufo calamita*) and the causes of these are not clear. Possibly degradation of the environment and overpopulation play a role here.

Mating and Breeding Behaviour

This is a very important aspect of amphibian behaviour and in many species it is the simplest to observe. The animals, which are otherwise very shy and secretive, come out *en masse* and whole populations concentrate in the breeding ponds during the normally brief mating period. The number of individuals in an area where traces of amphibians are sporadic during the rest of the year may be astonishingly large at such times. And the breeding individuals represent only part of the total number of sexually mature animals. Moreover, at the culminating point of the mating period most species are active both day and night and become much less secretive.

Some species (*Bufo viridis, B. calamita, Alytes* sp., *Bombina* sp.) go on breeding for several months (see pictograms for details). Sometimes rain, after a protracted period of drought, may provoke breeding behaviour. The same is true of the dredging of ditches or ponds after which species such as the Warty Newt (*Triturus cristatus*) may start to breed.

There is a marked correlation between breeding behaviour and migration. The animals move to the water, mostly that in which they were born. In this process relatively long distances are covered. The animals have to be present at the right time and both sexes have to find each other with the help of complex behaviour patterns. In this process most senses play a part and hence are well-developed (see the section on anatomy). Amphibian senses appear to react to certain chemical substances, including odours and pheromones, light, temperature and moisture. The ear is particularly sensitive to differences in atmospheric pressure, vibrations and sounds within particular frequencies that make up the mating call. The tactile sense, especially the tactile papillae on the surface of the breast and toes in males and the back in females, is also well-developed.

The olfactory sense appears to lead individuals to the right breeding spot in many species. Not always, however, since salamanders, newts and some frogs and toads always return to the pond where they were born, even if it has been filled in. Here geographical orientation or automatism plays a part.

Even animals unable to smell or see can orient themselves by automatism. During migration amphibians orient themselves towards one pre-determined pool. Having arrived they choose one particular spawning place within that pool. At times certain species (for example *Bufo bufo* and *Rana temporaria*) coincide in the time and exact location of their spawning grounds, and this results in a number of mistakes and in competition. It is mainly female frogs that incur casualties, some being poisoned or having their flanks torn by the sharp nuptial pads of the male *Bufo* sp. The function of this competition is not known.

In sufficiently large ponds individuals in different stages occur in different areas: egg-shedding couples, individuals that are not yet ready to mate and males in search of a mate along the banks. When an individual of the same species comes into the neighbourhood, then the male tries to grasp it (*Bufo, Rana*) or display before it (*Triturus*). If it is another male or a female that is not yet ready to mate, the result is a defence reaction or escape behaviour.

If this is not the case the result will be amplexus (the mating embrace), during which the male will keep off rival males by kicking at them (*Bufo, Rana*). See pictograms for the two types of amplexus. Salamanders and newts are led to their mates by pheromones (scent marks), followed by the typical display which in newts involves the crest and the vibration and lashing of the tail. Then the male emits a spermatophore (a compact package of spermatozoa) about 5mm in length, which is picked up by the female through her cloaca. The female then lays her eggs singly, attaching them to vegetation or rocks (see pictograms).

In frogs and toads females are attracted mainly by the croaking of males which leads them to their exact mating place in the pond. Observations suggest that the presence of certain algae is necessary for ovulation. Once ovulated the eggs have to be laid. Occasionally a female which has not arrived at the spawning place in time (*Bufo bufo*) lays her unfertilised eggs alone. In this case amplexus probably does not play any role in the ovulation. Should a pond have disappeared, the ovulation can be delayed until later the same year (in *Triturus* sp., *Bufo viridis, B. calamita*) or until the following year (*Bufo bufo*). While the eggs are being laid the female lowers her back (*Bufo, Hyla*), with the result that the male sits in a lower position and his hind limbs form a kind of bowl in which the eggs are fertilised. *Bombina, Pelobates* and

Peolodytes mate in loin amplexus and the female merely stretches her legs without lowering her back.

Pauses in egg-laying occur in all species. In frogs and toads the female displays a defensive reaction against the male who embraces her immediately after the eggs have been laid. *Bufo bufo* females make themselves appear very angular, since males are only stimulated by round, soft objects, even if these are not individuals of the same species. In the case of *Bombina* and *Rana* sp. females also produce a croaking sound.

Only Midwife Toads take care of their eggs: mating takes place on land and the male winds the strings of eggs around his hind limbs, carries them around for three weeks and moistens them until they hatch. Then he allows the larvae to swim out into a pool.

Homing Behaviour

Homing behaviour is well-developed in most amphibians and links up with migratory behaviour. Even outside the mating season the animals have an excellent sense of orientation. In some species this sense is based on automatism, but others have active orientation based either on the topography of the surrounding land, or on their position in regard to the sun or other celestial bodies.

Voice

Most male frogs and toads emit calls whilst salamanders and newts are silent. These calls are specific to each species thus preventing hybrid-isation (though this does occur in rare cases: *Bufo* sp., *Rana esculenta* complex). The mating call is the best known and some species only vocalise during that period. It is also the loudest call.

The function of the defence call has already been noted and usually it is rather faint.

In choruses the individuals often call alternately. The territory call serves to mark out and protect territory and is a manifestation of social behaviour. This can be observed best in *Hyla* during the mating season, *Bombina* and the Green Frogs (*Rana esculenta* complex). By means of this call individuals keep a minimum distance between one another.

The position of the territory within a pond is more important than its area. The formation of territories results in a well-balanced chorus which females find attractive. Once a couple has been formed it is not disturbed, except in the case of *Bufo bufo* and *Rana temporaria*.

The social position of territory-forming aquatic anurans is con-spicuous when they meet and during territory fights. The superior individual floats high in the water and is quite inflated, whereas the inferior individual lies deeper in the water, sometimes completely under it. In this position it is usually no longer troubled by its superior.

On being attacked an amphibian often emits a cry of distress. Tree frogs also emit a rain call.

Escape and Defence Reactions

When threatened by natural enemies amphibians usually try to escape by jumping away, backing off or keeping low. When a predator comes

HOW TO READ THE SYMBOLS

Physical characteristics

 Size of an adult individual (including tail)

(1) (2) (3) (4)

Shape of the eye pupil (frogs and toads only) Triangular (1), Vertical (2), Horizontal (3), Heart-shaped (4)

 Detached eggs deposited singly

 Small clutches

 Eggs of about 2mm, deposited singly or in clutches of 15 eggs

 Newt eggs, separate or in very small clutches, wrapped up in the leaf of a water plant

 Eggs normally on the back of a male

 Terrestrial species eyes directed sideways

 Aquatic species eyes directed upwards

 Female lays fertilised eggs singly

 Female gives birth to living larvae

 Female gives birth to fully developed and metamorphosed young

Habits and Behaviour

 Underside with virtually no markings

 Underside with markings

 Vocal species (most during the spring)

 Poison glands grouped into paratoid glands

 Male carries eggs on back until they hatch

 Solitary

 Gregarious

 Two vocal sacs on either side of the head

 One vocal sac under the head

 Axillary mating position

 Inguinal mating position

Egg Forms

 Can be found by day, in full sun

 Can be found only in cloudy or damp weather

 Relatively large clumps deposited without specific form

 Eggs are mostly black and are deposited in gelatinous strings

 Thick strings with more than one row, wound around plant stems

 Nocturnal species

 Nocturnal or in heavy rainfall by day

 Round Tails

 Tails flattened sideways

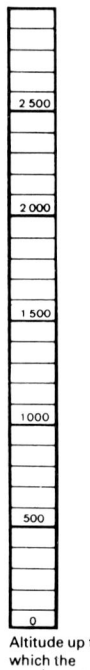

Column

1: period of activity in north-west Europe (the rest of the time the species hibernates)

2: breeding period in north-west Europe

3: period of activity in southern Europe (the rest of the time the species hibernates or aestivates)

4: breeding period in southern Europe

Range

Species that can be found far from water

2 500

2 000

1 500

1 000

500

0

Altitude up to which the species occurs

Habitat

Reed beds in still or running fresh water

Beaches, dunes, sandy soils

Near human habitations and urbanised areas

Ruins, walls, verges, churchyards, open quarries, sandpits (mostly with a pool)

Peat, marshlands, damp places in moorland

Deciduous forests

Pine forests

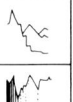

Hills, mountains, cliffs

Caverns, karst areas, caves, underground pools and streams

Pasture and arable land, hedgerows, plains, valleys

Damp forests in valleys and plains, alder-brakes

Meanderings of larger rivers, lakes and large ponds

Small water courses, brooks, running water

Small water surfaces, ditches, ponds, marshes, temporary pools, cattle watering places, still water

Artificially created waters

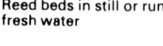

too close to *Bufo bufo,* it raises itself on all four legs, increases its size by inflation and makes straight for the attacker as if to hit it with its head. With snakes this trick seems to work, but on meeting a slow worm the toad may get confused. First it tries to snap at it as if it was prey, and if this produces no result the toad moves to its defence position. This happens because the slow worm is midway between the maximum size of the toad's prey and the minimum size of its predators.

Most salamanders, newts and toads exhibit a degree of thigmotaxis: when they wish to hide they feel safe only when in physical contact with a familiar refuge — a particular log, burrow, or root of a tree.

Other Reactions and Learning Ability

Most amphibian reactions and types of behaviour are instinctive. However, toads and frogs in particular can learn things. They can, for example, learn to keep away from bees, and can even tell on which day of the week their pond is most likely to be disturbed — usually Sundays. This can be proved by the fact that they behave differently on Bank Holidays! They can also learn to modify their behaviour during the year according to the food supply.

We still know very little about amphibians' learning abilities. For a long time they were underestimated since it was assumed that such primitive animals were not capable of learning because they lacked a cerebral cortex. This only shows that many researchers have not dealt with amphibians at first hand, and are not acquainted with their natural behaviour or environments.

Skipping actions occur when an individual is frustrated by two opposing stimuli. A similar sort of reaction is seen with *Bufo calamita* which detects its prey by smell. When it discovers the scent it looks for its source. But if it does not find its prey and its catching and swallowing reactions are checked, it suddenly yawns for no obvious reason.

Vestigial actions that once made sense can also be observed. A toad trying to swallow large prey will hunch its neck like someone facing a difficult task. Primitive Claw Frogs tear apart large prey using the claws on their hind limbs, and this is probably the origin of the toad's neck hunching.

FOOD

Adult amphibians are completely carnivorous and respond to moving prey only. Anything small enough to be swallowed which does not run, crawl or fly too rapidly may be considered as prey.

Salamanders and newts in their terrestrial stage eat worms, snails and the larvae of various insects. In the water the same animals eat insects and their larvae, mosquito larvae, several kinds of worm, water-fleas, small crawfish, molluscs, eggs of fish and other amphibians as well as their larvae. Sometimes cannibalism can be observed. Newt and salamander larvae have the same diet as their parents but their prey is smaller.

Frog and toad larvae are, on the contrary, herbivorous. They feed predominantly on algae, both free-living algae and algae growing on branches and other objects, which are chewed by means of a special rasp (rows of labial teeth). Amphibian larvae also live on vegetable and animal wastes.

The larvae of some amphibian species feed on the eggs and larvae of other species. In this way the Common Frog is able to counter competition from other amphibians which lay eggs later in the year in the same pond. Cases are known where no *Alytes obstetricans* larvae can be found in a pond, although the Midwife Toad laid eggs there, because of the presence of *Rana temporaria* larvae. As soon as the Common Frogs have metamorphosed in June and left the water, Midwife Toad larvae can be found again.

Apart from their usual diet of salamanders, adult frogs will eat several kinds of spiders, beetles and flies. Green Frogs leap out of the water to catch their prey and can be incredibly agile. Even dragonflies in flight can be caught.

Toads are more modest and restrict themselves to crawling insects. Often they prefer ants (*Bufo bufo, B. calamita*). Occasionally they eat flying insects too, but this is the speciality of Tree Frogs which are able to catch flying insects at night. They use their jaws or their sticky tongue, which can be flicked forwards. Ranidae, Bufonidae, *Chioglossa lusitanica* and *Hydromantes* sp. use the same technique. Other salamanders, newts, frogs and toads lack such tongues and have to catch their prey and grasp it with their jaws. When forcing their prey into their mouths both eyes and fore limbs may be used, and in some species small teeth prevent the prey from escaping.

Amphibians often have to fast for a very long time and are quite able to do so. During hibernation, when virtually all bodily functions are at a minimum, the animals do not eat. Nor do they eat during the breeding season or during poor climatological conditions when it is too cold, too hot or too dry. Yet their physiology is adjusted to living with little energy. Thanks to these qualities they have been able to survive where mammals and fish cannot and in situations where birds are forced to migrate.

NATURAL ENEMIES

Amphibians have numerous enemies which they combat by equally numerous adaptations. These include camouflaging marks and colours, their secretive and often nocturnal life, poison and warning markings (as in *Salamandra salamandra*) which attract the predator's attention, sometimes combined with conspicuous defence behaviour, most noticeable in the *Bombina* sp. of Fire-bellied Toads where the animal bends upwards and exposes its belly with its vivid red-black or yellow-black warning markings. As a last line of defence amphibians have numerous offspring every season (with some exceptions) — up to 10,000 eggs per brood for a single female.

Among mammals hedgehogs, rats, foxes and otters occasionally eat frogs. However the otter also keeps fish in check and these are a much greater threat to amphibians. The major threats come from carp, which eat all shelter vegetation, and carnivorous fish, including pike, trout and perch. These threats become particularly serious when these fish are introduced for angling purposes.

A number of birds eat amphibians, especially herons, kingfishers, storks, raptors and seagulls. Ducks and geese are also a threat, particularly the partly-domesticated individuals which are very common in many European wetlands, especially ponds in parks.

Among reptiles there are several specialised predators of amphibians and their larvae, notably the grass snake, dice snake and viperine snake, and to a lesser extent the European terrapins which go after larvae and newts in particular.

Among amphibians themselves the recently introduced American Bull Frog (*Rana catesbeiana*) has proved to be a serious predator of other amphibians, especially in northern Italy where it was first introduced to Europe. Green Frogs too eat other amphibians and Warty Newts (*Triturus cristatus* and *T. carnifex*) feed on large numbers of amphibian larvae including their own species.

Amphibians are regularly infested by several kinds of invertebrate parasites including bacteria, insects, flagellatae, worms such as *Pseudomenos* sp., *Eimeria* sp., *Cestoda, Nematoda, Trematoda, Trypanosoma* sp. and molluscs, which infest the gills (*Oodinium, Trichodina*).

In addition there are specialised and lethal parasites such as *Lucilla bufonivora* flies which lay their eggs in hot summers, especially in northwest Europe, in the nostrils of Common Toads. The maggots devour their host alive until the animal dies, having first displayed totally abnormal behaviour. Up to 30 per cent of the adult population may be affected. This phenomenon and its regulatory function has not yet been fully examined.

These natural predators and parasites cannot be regarded as real threats: both amphibians and their predators are elements in one natural whole. Predators have a selecting and regulatory function among amphibian populations so that only the fittest and best-adapted individuals survive and breed. The situation is different when man starts to interfere with the balance of nature.

Under natural conditions amphibians are able to combat several kinds of dangers and have adapted themselves to do so throughout their evolution. Only when amphibians are threatened by human activities, which drastically upset the natural balance, can natural predators at times prove fatal. Specialised predators such as the grass snake often perish themselves soon after the amphibian populations have disappeared.

ENDANGERED SPECIES

In the last 20 years a sharp decline in all amphibian species has been observed. Of the 48 species left in Europe, 14 are on the list of the *IUCN Red Data Book* (Vol III *Amphibia/Reptilia* by R. E. Honegger). In West Germany 63.2% of native amphibians are in danger (Blab & Nowak). At least 7 of the 17 Swiss species have shown a marked decline (K. Grossenbacher, R. Honegger). In Italy (S. Frisenda, F. Petretti, M. Capula), France (J. Fretey, Baumgart and others) and Spain (A. Salvador) there are large regional and local differences. Although no exact data on the whole of these countries is available, it is clear that in general amphibian numbers are being reduced, especially *Pelobates, Pelodytes, Hyla* and *Bombina* species. Britain has only 6 native species, one of which is seriously endangered, while 4 other species are declining (I. F. Spellerberg). In the Benelux countries 5 out of their 15 species are about to become extinct, with the remainder declining sharply (A. Stumpel, P. De Fonseca, D. Ballasina, C. H. Parent).

This situation is markedly worse than that of mammals, with 12% of species endangered in Europe, and birds with only 10%.

Natural Factors

In addition to natural enemies, there are other natural factors capable of causing a decline.

Climatological factors

In the whole of western Europe the summers appear to have become cooler and the shortage of sunshine noticed by Jackson in the UK between 1960 and 1970 has certainly affected the most thermophile species at the northernmost part of their range (*Hyla, Bombina, Alytes, Rana esculenta* complex, *Triturus cristatus*). The decrease in rainfall in recent years will not have helped the equally decreasing number of breeding ponds (Parent, 1983). Cold and drought take their toll of victims too, although they are only temporary fluctuations and extreme climatological conditions help to eliminate the weakest individuals, thus creating a stronger and better adapted population.

Other natural factors

The choking of ponds by increased vegetation can reduce a local population and the same is true of denser growth in surrounding shrubs and trees, resulting in less sunshine reaching the pond or summer habitat of thermophile amphibians (*Pelobates, Alytes, Pelodytes, Hyla*).

However, neither natural predators nor other factors can — under normal conditions — result in a species disappearing from an area, although climatological changes should not be underestimated. So when a species does die out in certain areas, as now happens regularly in Europe, this is nearly always the result of an imbalance or change in the habitat caused by human activity. Europe has such a large number of small relic populations of amphibians that many of them may no longer be able to adapt to environmental or climatological changes.

Human Threats
Loss of habitats

Human threats may result in the total or partial loss of amphibian habitats. Populations can survive successfully only when their summer, winter and breeding habitats remain intact, as well as possible migration routes between these habitats.

A considerable number of breeding ponds have disappeared — most of them during the last 20 to 30 years. This is the result of drainage, due to the expansion of agriculture and the fight against mosquitoes;

dumping, due to urbanisation, road-building, rubbish tips and allotments; and drinking pools for livestock being replaced by concrete water-troughs.

Many ponds have dried out due to the lowering of the water level to create canals and reservoirs. Temporary pools — such as cattle drinking pools or ponds in woods or dunes — are especially vulnerable. The regularisation of streams that used to flood and leave a series of pools has been another major factor in the disappearance of breeding ponds.

Habitats such as ponds, pools in peat-digging areas, flooded sandpits and quarries, castle moats, etc, have become rare. Springs are drawn off directly without allowing water to flow freely. This endangers the breeding of newts and salamanders, and even if the animals manage to get into the water tanks, these usually prove to be lethal traps from which there is no escape. It goes without saying that the re-routing of a river bed has equally unfortunate consequences.

The decreasing number of suitable breeding ponds results in a concentration of amphibians in the remaining areas, causing increased competition and predation among and even within species (Brockelman 1969, Heusser 1970, Parent 1983, De Fonseca). The rarest species are the first to disappear.

Summer habitats are also being reduced drastically: heathlands, ruins, dune areas, wetlands and natural forests are becoming increasingly rare in Europe. Quite a number of habitats are destroyed by the establishment of holiday villages and other recreational sites. The construction of weirs and large-scale reforestation, especially with foreign conifers, has caused many habitats to change or disappear.

Changing of habitats: damage and pollution

In many areas there seem to be enough natural habitats left, yet nonetheless amphibians are declining. This is caused by external factors working slowly throughout the whole ecosystem. Amphibians are essential components of the ecosystems in which they live. They are reliable indicators in determining the health of the environment or its degree of damage and pollution.

During the last few years quite a number of changes — unfortunately irreversible — have taken place in the landscape. The variety of landscape has decreased and only a few natural areas are left. Usually they are very small and isolated and are substantially less valuable than before.

Apart from causing pools and other habitats to disappear, large-scale intensive agriculture has a very negative impact on the remaining habitats. This is caused by spraying herbicides, destroying among other things important water flora and affecting the mucous covering of the amphibian skin so that it becomes susceptible to mycoses.

The spraying of insecticides has disastrous consequences including lethal poisoning and deformed larvae. Many pools have been rendered uninhabitable as a result of being used to rinse out spraying equipment. The splendid area of Neusiedlersee (Burgenland, Austria, Hungary) has lost entire populations of amphibians owing to such practices, especially among vine-growers.

Over-manuring with fertilisers which are rich in phosphates causes eutrophication of the water and the rapid destruction of ponds. In some cases it also leads to a spectacular growth of algae accompanied by acute oxygen deficiency and the death of amphibian larvae. In most cases the consequences of fertilisers are only noticeable after a long time. The enrichment of the soil causes a gradual change in vegetation, which in its turn changes the insect fauna and hence the amphibian fauna.

Water pollution by dumping household waste (detergents) and industrial waste (chemicals, oil, cellulose, heavy metals, etc) has had fatal consequences. The same is true of salts from factories, mines or road salt, used in winter to thaw ice — all of which end up in the water.

Noise from railways and motorways is another form of environmental pollution with fatal consequences for amphibians. Certain sound frequencies prevent individuals from hearing mating calls. They became disorientated and breeding cannot take place. In one case a population of Tree Frogs died out after the construction of a motorway, though their habitat had remained intact (De Fonseca).

Some ponds are destroyed because they were situated in a pasture that has been changed into arable land even if the pond itself was not filled in. In larger waters the agitation of the water caused by water sports may cause a decrease in the penetration of light into the water, resulting in reduced aquatic vegetation, to which many amphibians attach their eggs, and reduced water quality.

Overall the quality of water is decreasing throughout Europe. Not only surface water but underground water too. Three-quarters of the 600 caves in Belgium are polluted (Parent, 1983). In north-east Italy and Yugoslavia this pollution will soon cause the final extinction of the Olm (*Proteus anguinus*) (R. Honegger, 1980; M. Capula, 1982).

Pollution even stretches over borders. In frogs from Turkey, imported into Switzerland for consumption, analyses found up to 0.2ppm of mercury and 0.15ppm of lead (R. Honegger, 1977). Throughout Europe acid rain renders many spawning places unsuitable.

Isolation of habitats

Due to the high density of human population, increasing urbanisation and the dense network of roads, there are virtually no large coherent natural areas left in Europe. The remaining countryside consists of dispersed islands in the surrounding cultivated land. Usually they are not interconnected, thereby increasing the risk of extinction. Unlike birds which can fly from one suitable place to another, amphibians are relatively immobile animals.

Should a population suddenly disappear in one of the few connected areas, then immigration from another population is possible. However, this is clearly impossible with isolated areas and hence Europe contains many empty amphibian habitats. This happens more frequently in certain species than in others. Changes in habitat constitute a greater threat when endangered species are less able to adapt rapidly and efficiently.

In spring many amphibians migrate to their mating places (*Bufo bufo, B. viridis, B. calamita, Rana temporaria, Triturus* sp. and others).

When the hibernation area is separated from the spawning place by a motorway, individuals are run over in large numbers. This sad phenomenon is best known in the Common Toad (*Bufo bufo*). Each year thousands of individuals perish before they are able to breed. In this way entire populations of amphibians can be exterminated within a few years.

Introduction of other animals

It has been proved that the introduction of native and foreign amphibians and other animals, usually with commercial objectives, has a negative impact on natural populations. Even the introduction of the same species — but from a totally different place — may alter the genetic patterns of a population, especially when it is a small population that has been isolated for a relatively long time and could therefore constitute interesting matter for scientific study.

The introduction of closely related species may cause an entire population to die out. In Europe *Triturus cristatus* (Warty Newt) has interbred with *Triturus marmoratus* (Marbled Newt), producing *T. x blasii* (*T.m* female x *T.c* male) or *T. x trouessarti* (*T.c* female x *T.m* male). The offspring are sterile but can breed with one or other of the parent species. Similar introductions have led to the hybridisation of the *Bombina* species (*B. variegata, B. bombina, B. orientalis*). Green Frogs are often commercially introduced for frog's legs and escaped individuals (usually *Rana ridibunda*) may alter native populations and cause them to die out. The introduction of the American Bull Frog has already been mentioned. Other introduced aliens include *Rana dalmatina* (Agile Frog), *Rana graeca* (Stream Frog) and *Xenopus laevis*. The latter is originally from southern Africa and is often used here for genetic experiments in laboratories and pregnancy tests on women.

The dangers connected with the introduction of fish, birds and mammals have already been mentioned. The introduction of small viviparous topminnows such as the *Gambusia* species from America has already exterminated several salamander subspecies in Venice and Sardinia. With mammals it is cats that have the most serious impact.

Capture and destruction of amphibians

In former times amphibians were hunted and occasionally inquisitive children still catch newts, though this does no harm. Far more serious is the capture of millions of frogs each year for the sake of gourmets' frog's legs.

In addition the pet trade deals in increasing numbers of amphibians — especially rare ones. Other giant clients are hospitals, universities and even schools, wanting to use frogs or toads for dissection. Nearly every biologist, veterinarian or physician will have dissected a few frogs during his training. Myriads of amphibians taken from seriously endangered environments are used for scientific research.

It is estimated that in the French-speaking area of Belgium alone at least one million Common Frogs (*Rana temporaria*) a year are caught to be eaten, so that this widely distributed species has become rare (Parent, 1983). The Common Toad (*Bufo bufo*) is now a substitute.

31

From 1 July to 31 December 1975, up to 136,386lb of Tree Frogs (*Hyla arborea*) were imported from Turkey, Bulgaria, Hungary and Yugoslavia into Switzerland. The average weight of an individual being 2.82oz, this represents at least one million animals (Honegger, 1977). In 1976 Greece exported 2.2 million Tree Frogs for consumption, and from 1968 to 1970 Italy exported over 47 million frogs and toads for consumption, the pet trade and research. Within the same period of time Italy exported at least 96,000 salamanders and newts (Bruno, 1973). In some parts of Poland the Green Frog (*Rana esculenta* complex) runs the risk of complete extinction through capture for export (Kowalski, 1969).

It is clear, therefore, that many smaller and often isolated amphibian populations can be destroyed by such captures within a short time, especially when this takes place during the mating season. Males can easily be found then by listening for their calls (*Hyla arborea, H. meridionalis, Bufo viridis, B. calamita*). Many species that are otherwise hard to find, occur in great concentrations during the mating season.

Capture by terrarium enthusiasts should not be underestimated either. Throughout Europe there are examples of populations that have been destroyed by a handful of enthusiasts.

PLATES

The 48 species pictured in the colour plates represent all known European amphibians. For easier identification more than one photograph is provided for some species. Unless otherwise stated the pictures are of a male individual.

The photographs show typical animals. Here and there some frequently recurring variations are given, such as nuptial dress or variations in markings, and for a few species additional photographs show the striking colours of their undersides.

IDENTIFICATION OF AMPHIBIAN LARVAE

Pictures of juvenile animals are indispensable in a field guide to amphibians. Quite often the presence of larvae is the only indication that a species inhabits a given area. This is particularly true of *Pelobates* sp. (Spadefoots) and some salamanders and newts.

Most tadpoles and salamander and newt larvae are easy to find in ponds, streams and ditches, whereas adult individuals are often much more difficult because of their secretiveness outside the mating season and their avoidance of daylight.

Pages 34–5 show the most easily identified types of larvae. In a number of cases divisions into species has not been possible and characteristics are given only for the genus.

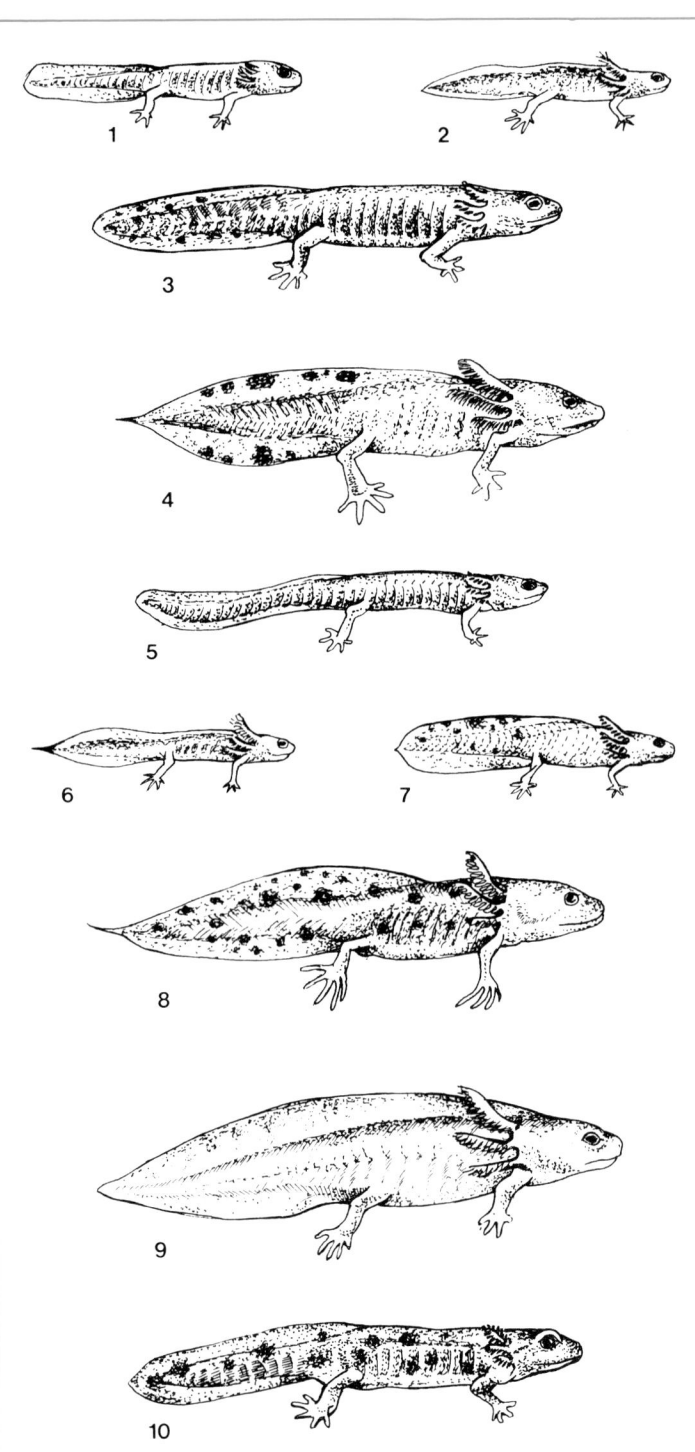

Salamanders and Newts (Caudata)

 1 *Salamandrina terdigitata* (Spectacled Salamander)
 2 *Triturus vulgaris* (Smooth Newt)
 3 *Euproctus* sp. (Brook Salamanders)
 4 *Triturus marmoratus* (Marbled Newt)
 5 *Chioglossa lusitanica* (Golden-striped Salamander)
 6 *Triturus helveticus* (Palmate Newt)
 7 *Triturus alpestris* (Alpine Newt)
 8 *Triturus cristatus* (Warty Newt)
 9 *Pleurodeles waltl* (Sharp-ribbed Salamander)
10 *Salamandra* sp. (Salamanders)

Frogs and Toads (Anura)

11 *Pelobates* sp. (Spadefoots)
12 *Rana catesbeiana* (American Bull Frog)
13 *Discoglossus* sp. (Painted Frogs)
14 *Alytes* sp. (Midwife Toads)
15 *Bufo bufo* (Common Toad)
16 *Bufo viridis* (Green Toad)
17 *Pelodytes punctatus* (Parsley Frog)
18 *Rana esculenta* complex (Green Frogs)
19 *Rana temporaria* (Common Frog)
20 *Hyla* sp. (Tree Frogs)
21 *Bufo calamita* (Natterjack)
22 *Bombina* sp. (Fire-bellied Toads)
23 *Rana dalmatina* (Agile Frog)

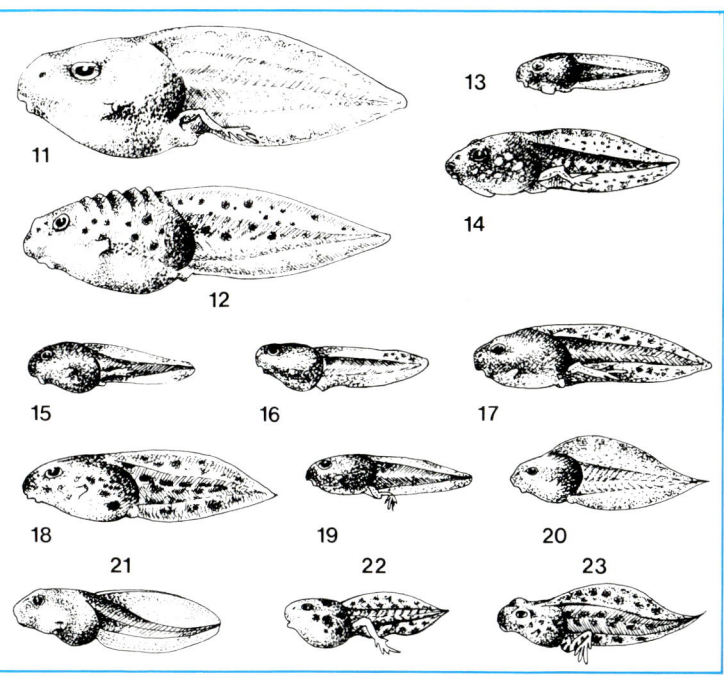

HOW TO USE THIS GUIDE

The species are divided into three groups:
1 Salamanders and newts (1–21), with orange top corner of page
2 Toads (22–35), with brown top corner of page
3 Frogs (36–48), with blue top corner of page

The latter two categories are arbitrary divisions which are not strictly scientific. They should, however, make identification simpler.

To find the exact species you should start with the colour code (right-hand top corner of the page), the colour pictures and then the data in the pictograms.

The pictogram data gives at-a-glance details of distribution and habitat plus typical characteristics which are conspicuous during the mating season, especially egg-laying sites, characteristic mating habits, the different types of eggs, and the breeding and activity cycles, with separate notes for north-west and southern Europe. These characteristics have been given special attention because during the mating season amphibians are easier to observe.

If the data is intermediate between two symbols, the first and most important is completely coloured, and the second only partly.

The pictograms have been kept as simple and practical as possible. To make certain, they have been tested on adults and children without any special knowledge of amphibians. Even without any explanation, they understood at least 8 out of 10 symbols.

The characteristics that could not be put in the pictograms are mentioned separately. Finally, a separate table indicates to what extent each species is endangered in European countries:

B	=	Belgium	D	=	West Germany
NL	=	The Netherlands	I	=	Italy
L	=	Luxembourg	E	=	Spain
GB	=	Great Britain	CH	=	Switzerland
F	=	France	A	=	Austria

=	species maintains itself with virtually no decline
✔	species is declining in numbers and in number of places found
✔	species is declining rapidly
♭	species has nearly become extinct in the country in question
†	species extinct

SALAMANDERS AND NEWTS

Salamandridae
Plethodontidae
Proteidae

Typical habitat

1 **Salamandra salamandra**

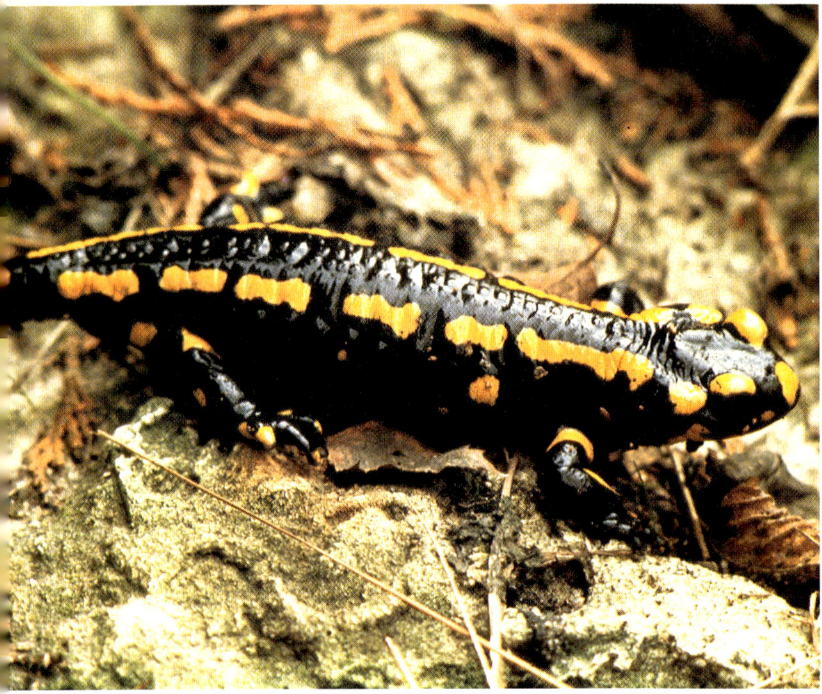

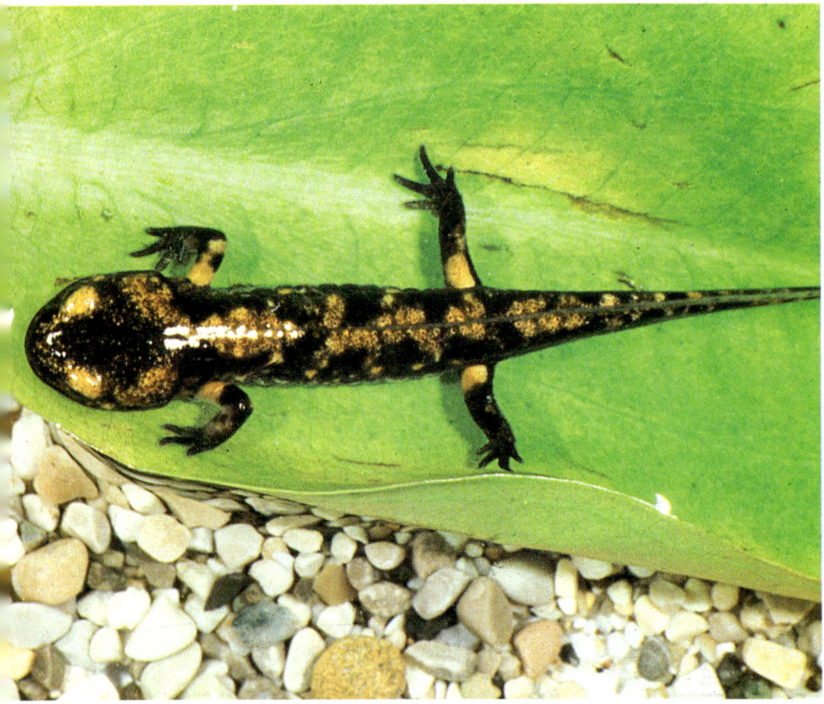

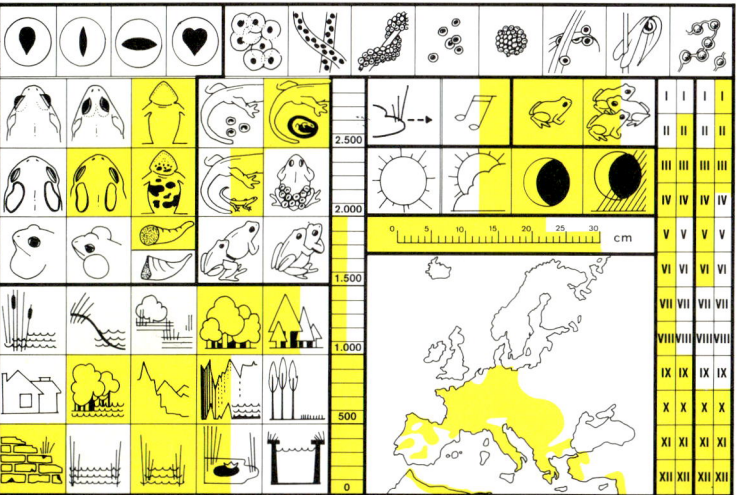

Fire Salamander

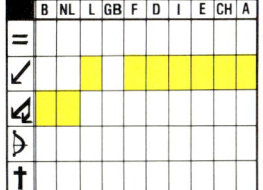

	B	NL	L	GB	F	D	I	E	CH	A
=										
↙										
◁										
⋺										
†										

Pattern and colour highly variable, from spotted to striped and from yellow to red. Reaches an age of at least 20 years. Quite poisonous, though not dangerous to man. Gives birth to 10–15 larvae. Has a faint mating call resembling that of *Bufo bufo* and sometimes emits a sound when grabbed. The photograph above shows the typical form. On the left-hand page (top) is the western striped form (*Salamandra salamandra terrestris*) and (bottom) the larva just before metamorphosis.

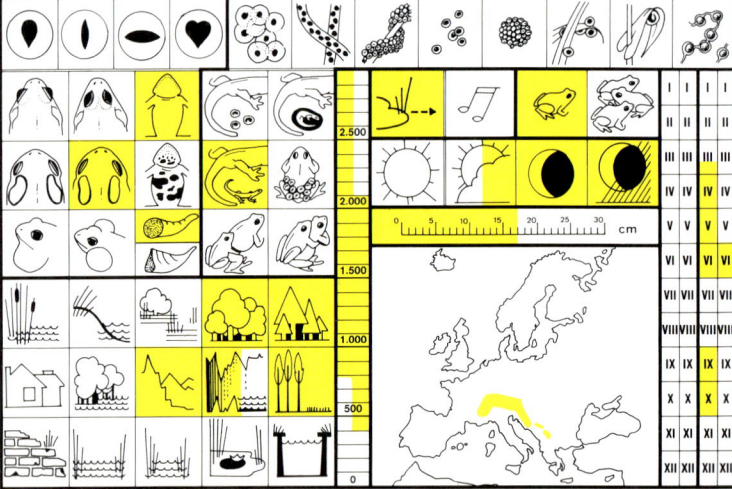

Alpine Salamander

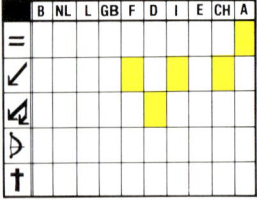

	B	NL	L	GB	F	D	I	E	CH	A
=										
↙										
⬙										
◁										
✝										

Locally can be quite common. Mostly 2–4 young after a two to three year gestation period. Colour sometimes brownish.

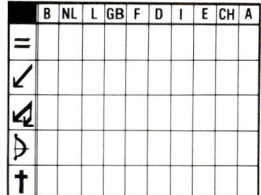

Luschan's Salamander

	B	NL	L	GB	F	D	I	E	CH	A
=										
↙										
↙										
▷										
†										

Back and throat covered with small prickles. Little-known species. Moves rather quickly for a salamander.

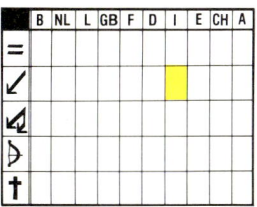

Spectacled Salamander

	B	NL	L	GB	F	D	I	E	CH	A
=										
↙										
◁										
▷										
†										

Only European salamander species with four toes on its hind legs. Mates on land. Feigns death when threatened or curls up coloured underside of tail.

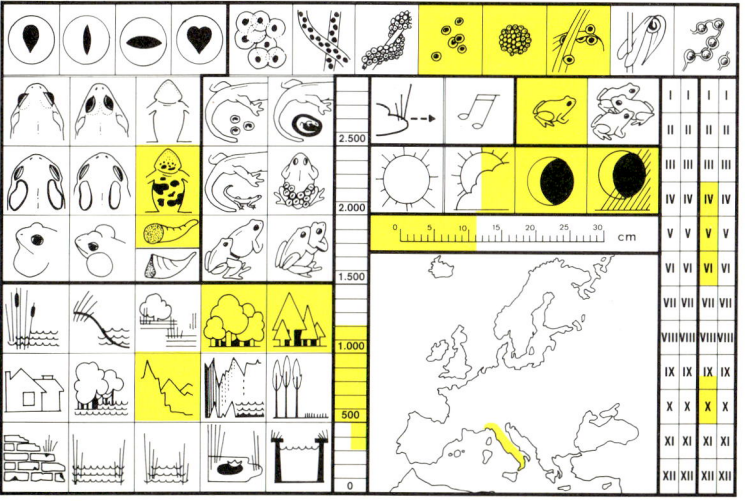

Chioglossa lusitanica

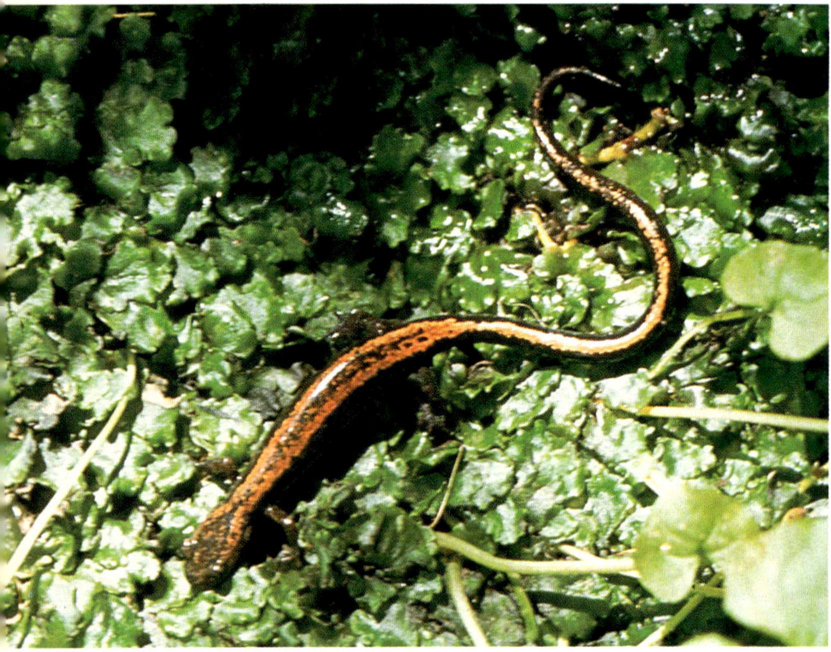

Golden-striped Salamander

	B	NL	L	GB	F	D	I	E	CH	A
=										
↙										
⇙										
⊳										
†										

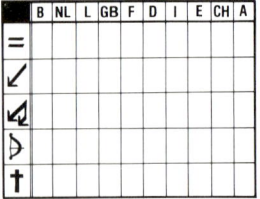

Tail nearly twice as long as body. Can shed its tail as lizards do (autotomy), then grows a grey tail. Catches food with its tongue. Lungs reduced in size.

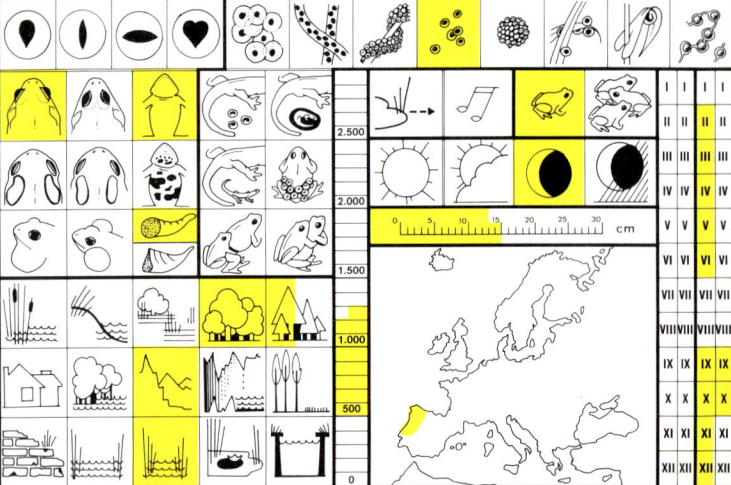

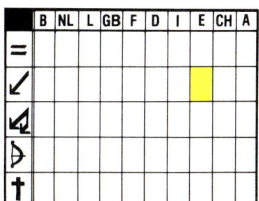

Sharp-ribbed Salamander

	B	NL	L	GB	F	D	I	E	CH	A
=										
↙										
◁										
⏋										
†										

Largest European salamander. Very strong swimmer. Has sharp rib-tips, sometimes projecting through the body. Eggs small, enveloped by a thick gelatinous layer.

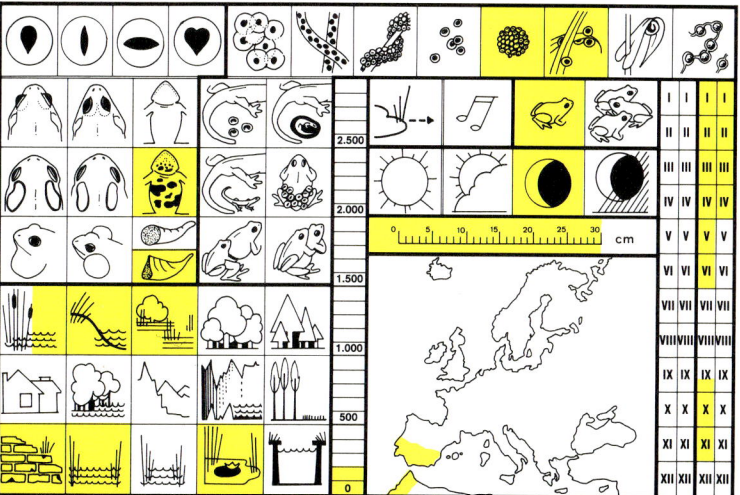

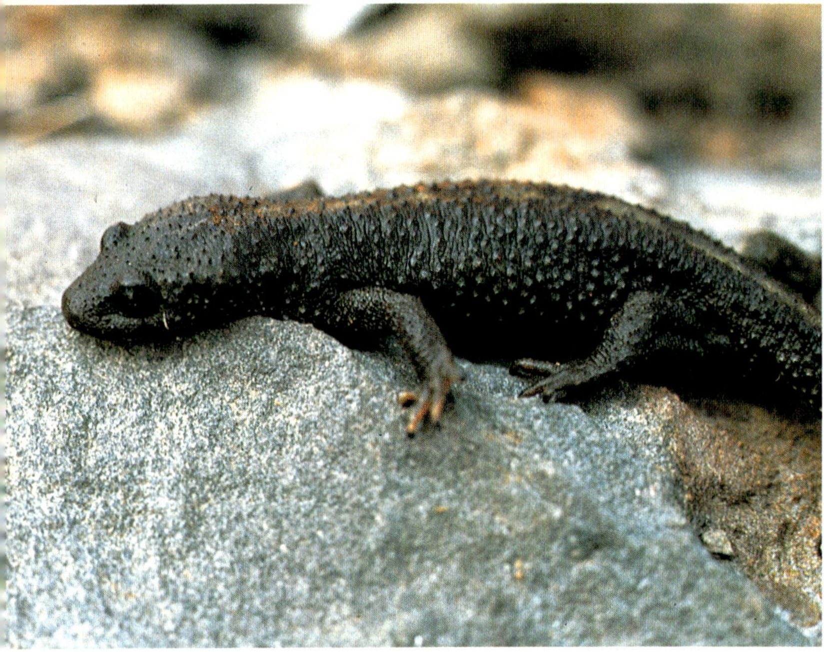

Pyrenean Brook Salamander

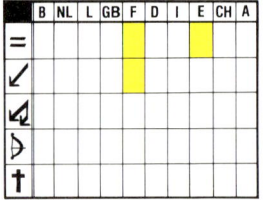

Rough skin. Local size variations. Can be found in water below 15°C. Never swims directly to the surface as *Triturus* species do. Eggs relatively large, but few in number. Fingers and toes of larvae have horny tips.

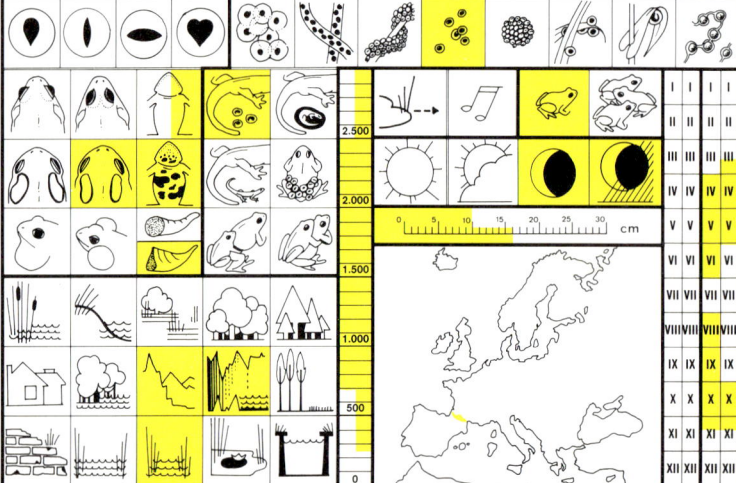

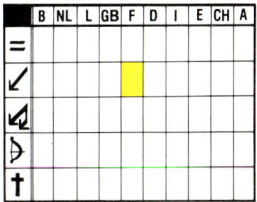

Corsican Brook Salamander

	B	NL	L	GB	F	D	I	E	CH	A
=										
✓					▨					
◁										
▷										
†										

Has smoother skin than *E. asper*. Can also be greenish.
No lungs.

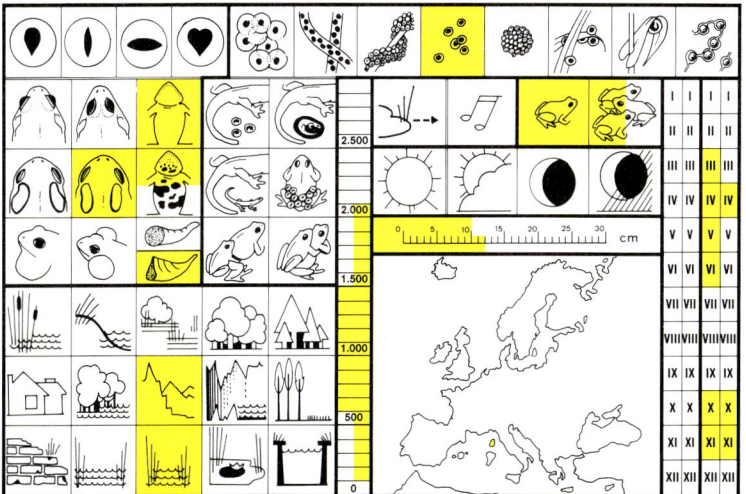

9 Euproctus platycephalus

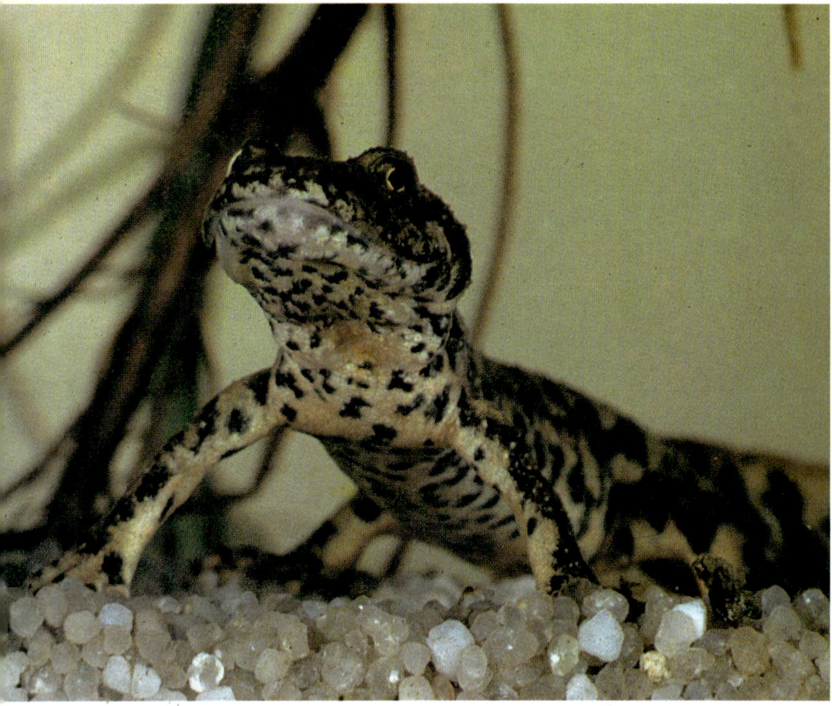

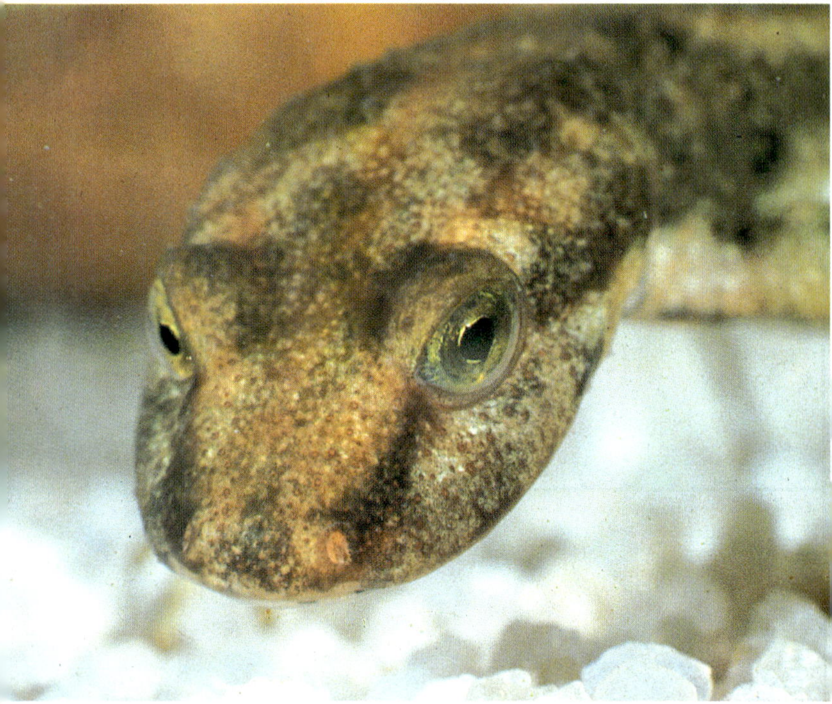

Sardinian Brook Salamander

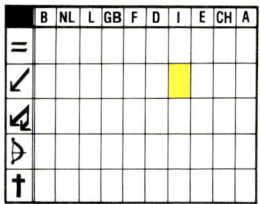

Resembles *E. montanus*, but head broader and flatter. Size between that of both other *Euproctus* species. Always brown.

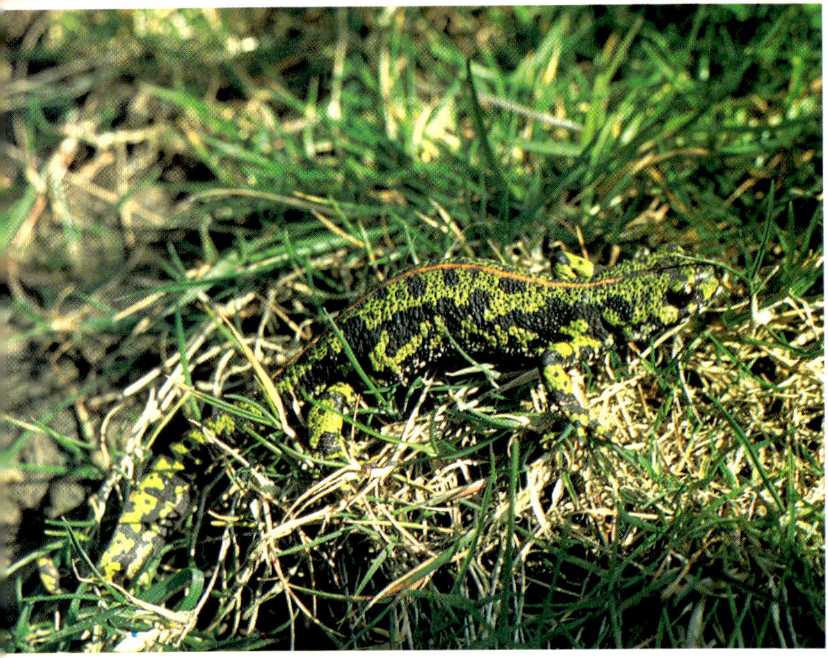

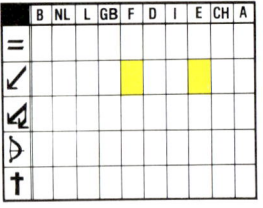

Marbled Newt

	B	NL	L	GB	F	D	I	E	CH	A
=										
↙					■			■		
↙										
⋺										
†										

Males in spring with large crest. Females with yellow stripe on back. Lays 200 to 380 eggs. Can be found in water with abundant vegetation during mating season. May also hybridise with *T. cristatus* under natural conditions.

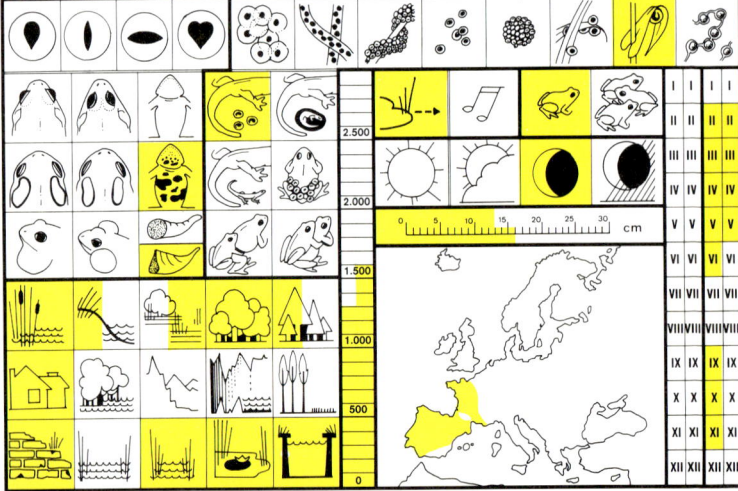

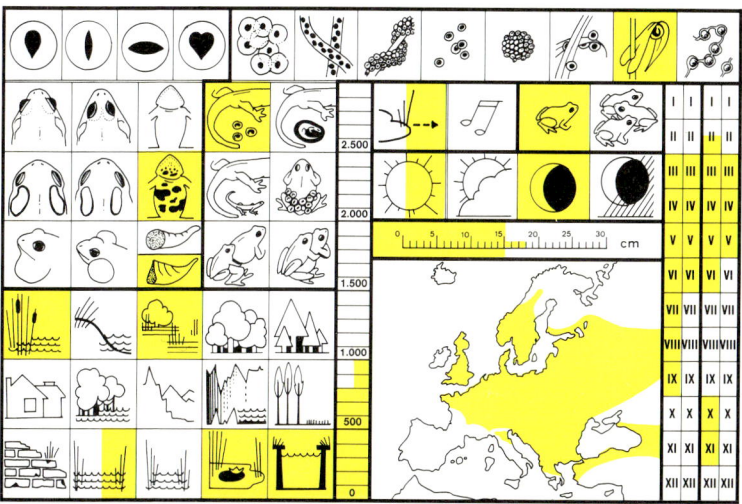

Warty Newt

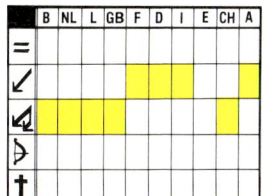

	B	NL	L	GB	F	D	I	E	CH	A
=										
↙										
⋺										
†										

Male often with large black crest and white dots. Mostly in water with abundant vegetation. Males fight for territory. Female often larger than male. Quite voracious predator of eggs and larvae of other amphibians. Second largest newt in Europe. Lays 200–300 eggs. More sensitive to pollution than other newts. *T. cristatus dobrogicus* may be separate species.

Triturus carnifex

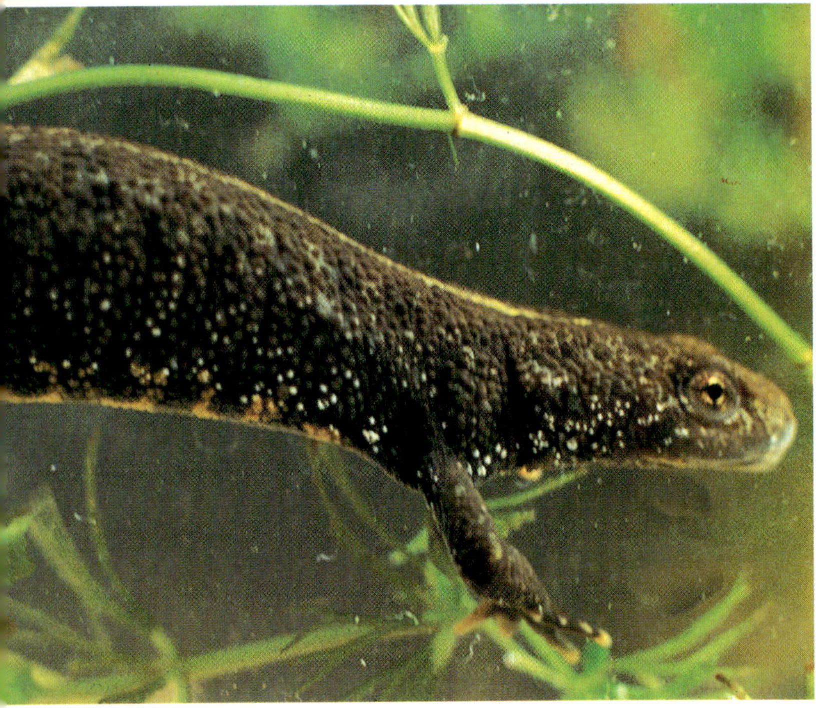

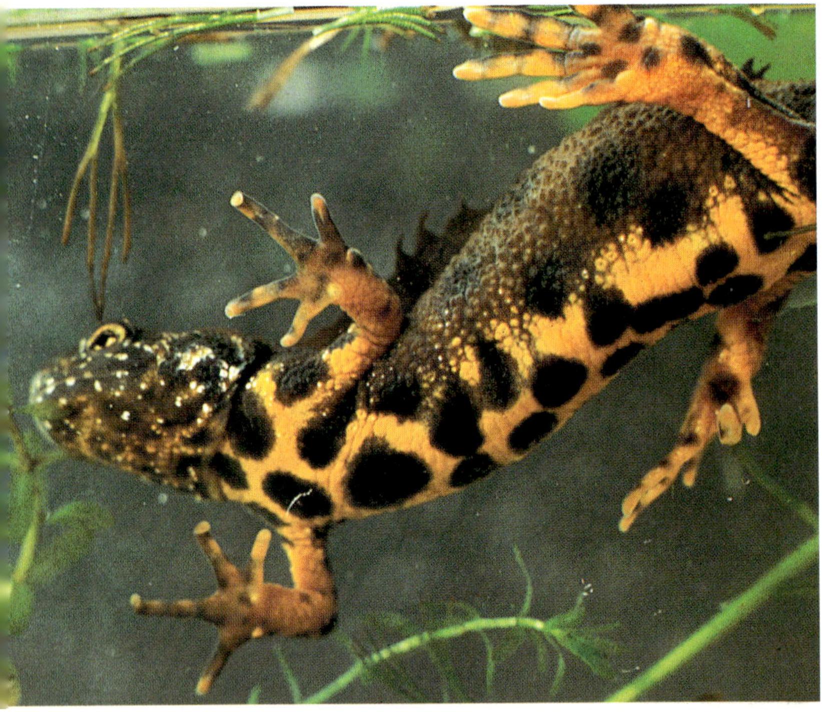

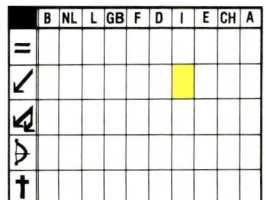

Italian Warty Newt

	B	NL	L	GB	F	D	I	E	CH	A
=										
↙										
◢										
▷										
†										

Strong resemblance to *T. cristatus* but less slender and more brownish. Recently recognised as a separate species. Female is shown top left, with the male below.

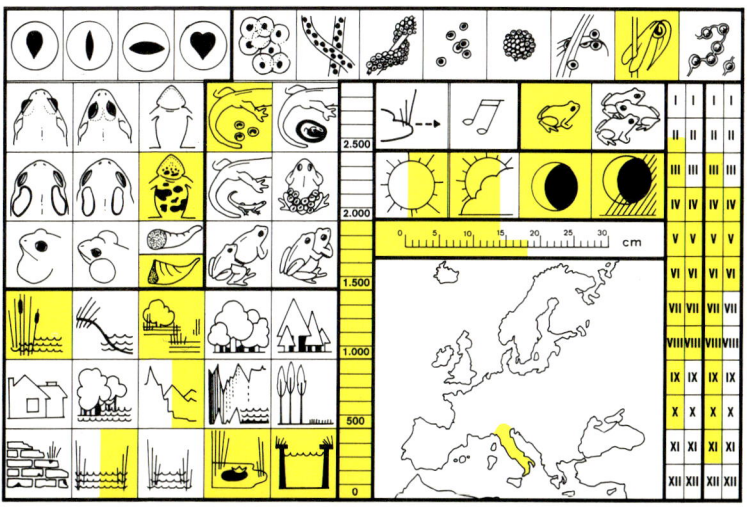

 Triturus alpestris

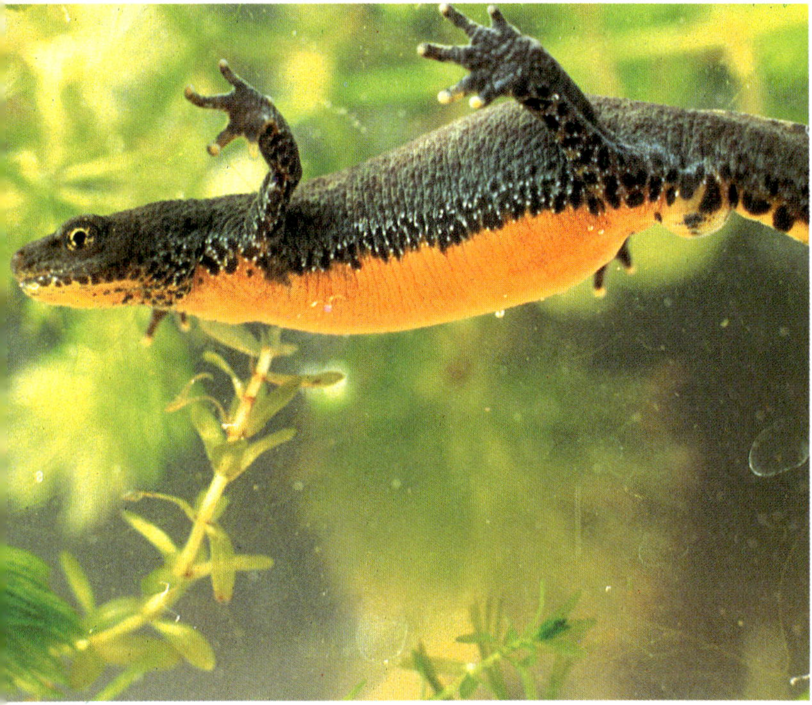

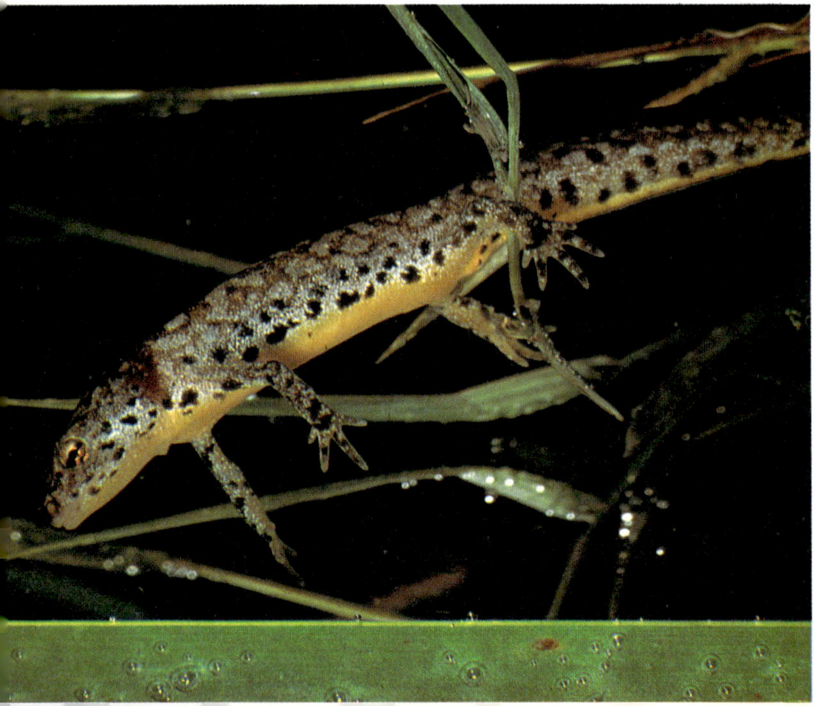

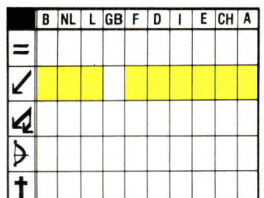

Alpine Newt

	B	NL	L	GB	F	D	I	E	CH	A
=										
↙										
↙										
▷										
†										

Not restricted to Alps. Male with very small crest in spring. Bluish, but female rather olive. Neoteny occurs, especially in Yugoslavia. Can remain in water throughout the year. Male is shown above, with the female top left and the juvenile below.

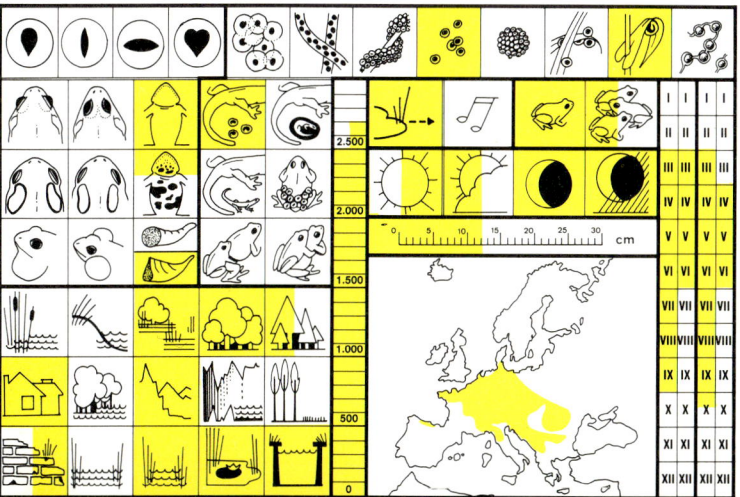

 Triturus vulgaris

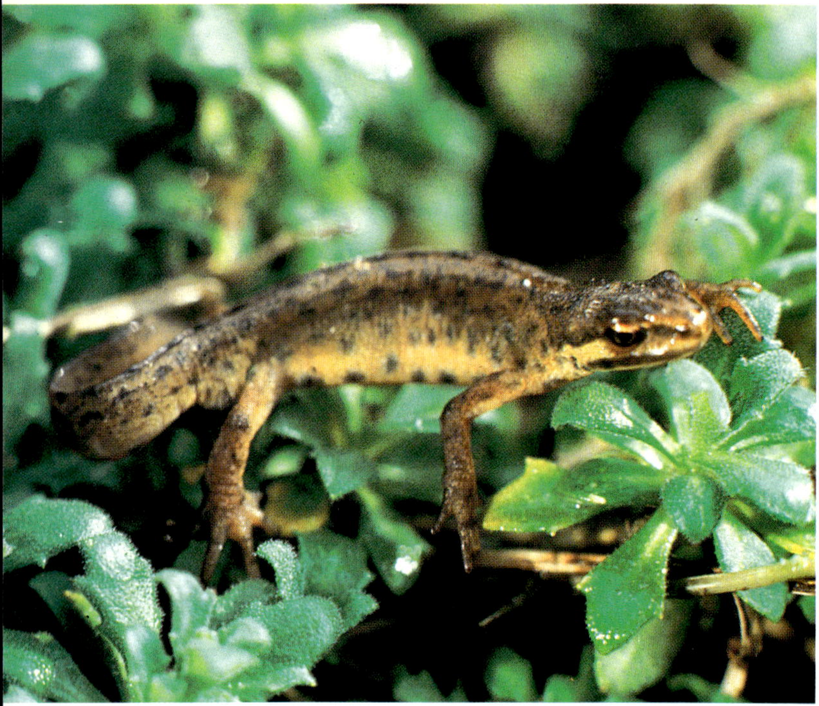

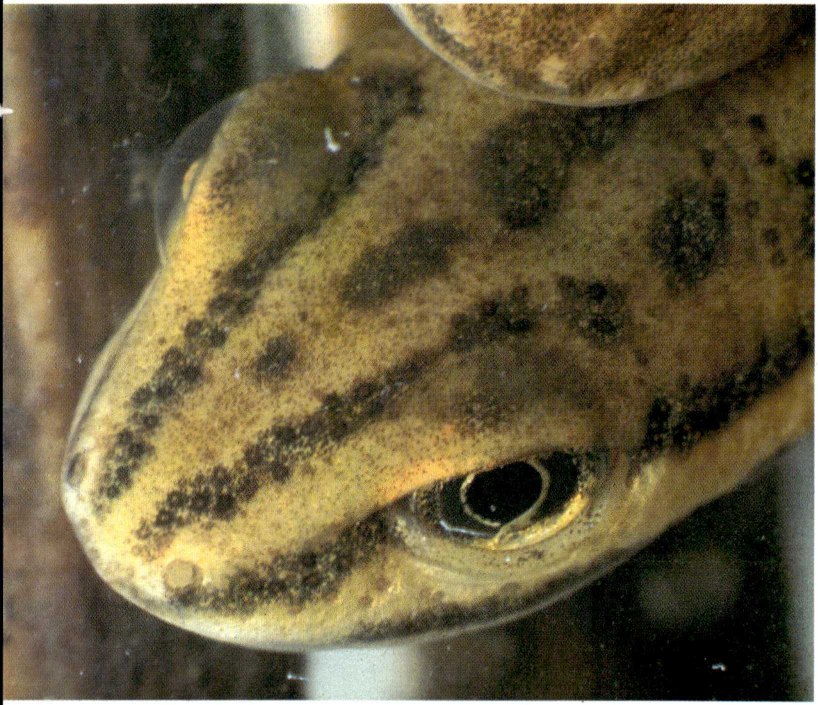

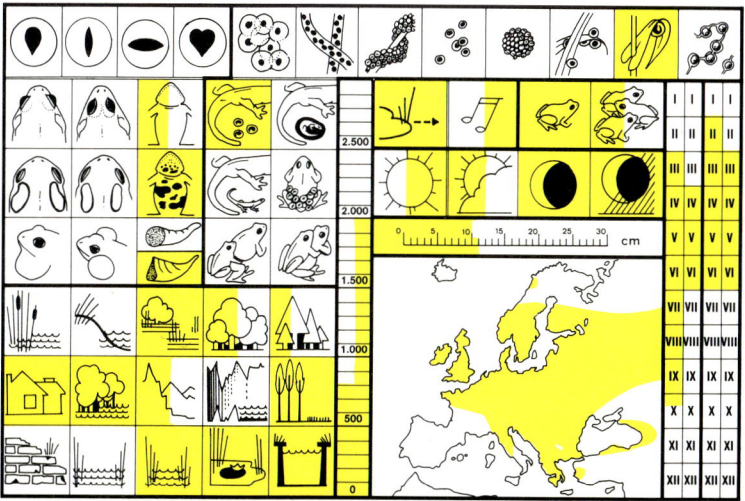

Smooth Newt

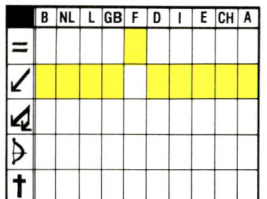

	B	NL	L	GB	F	D	I	E	CH	A
=										
↙										
⬠										
▷										
†										

Widely distributed. Sometimes occurs in polluted water. Male has a large crest in spring. Female difficult to distinguish from female *T. helveticus*. *T. vulgaris* often has spots on throat, *T. helveticus* has none. Lays 100–350 eggs. The terrestrial form is shown top left and the underside of the female overleaf.

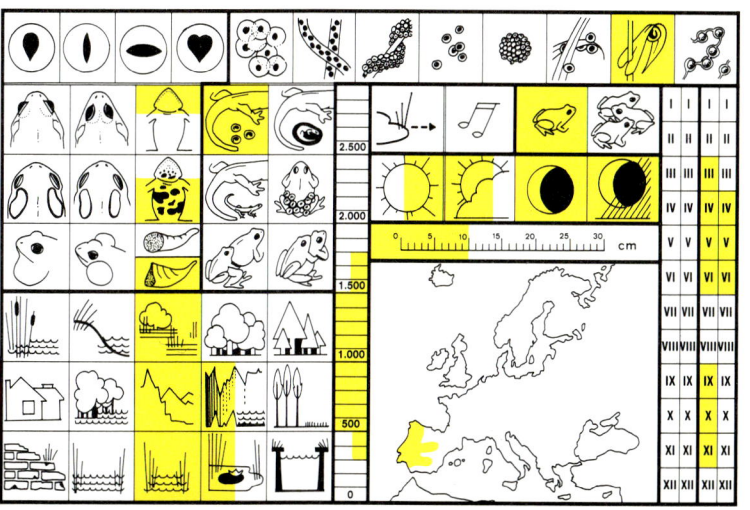

Bosca's Newt

	B	NL	L	GB	F	D	I	E	CH	A
=										
↙										
◤										
⇾										
†										

One of the smallest European Newts. Virtually no difference between male and female.

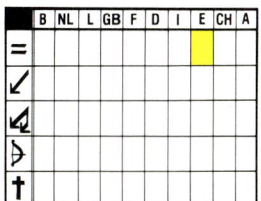

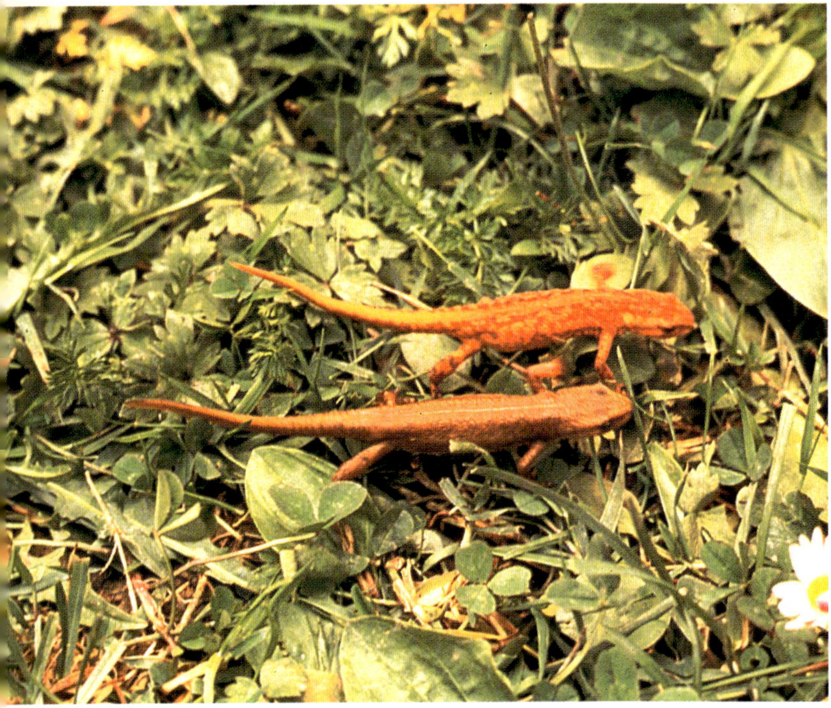

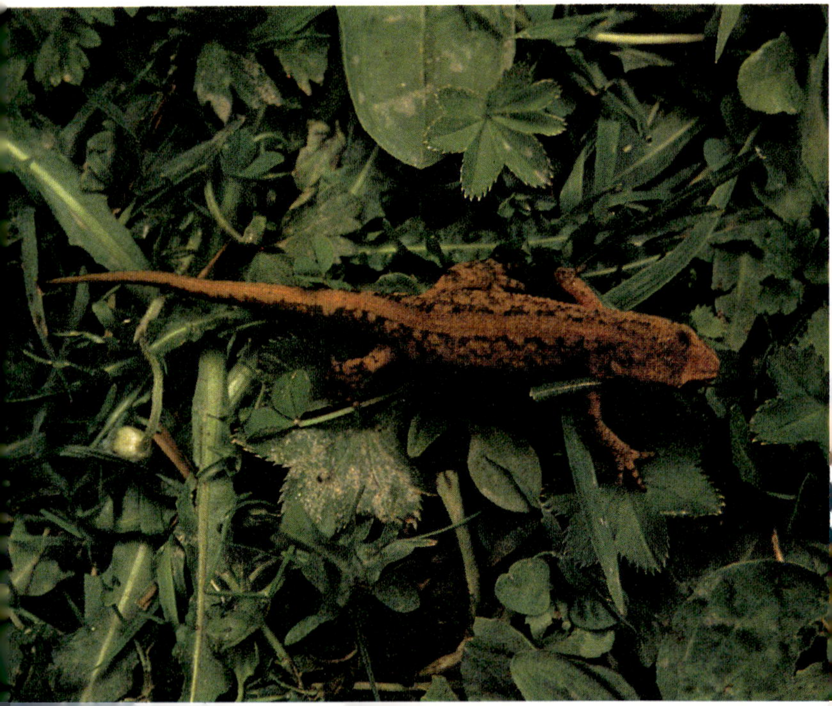

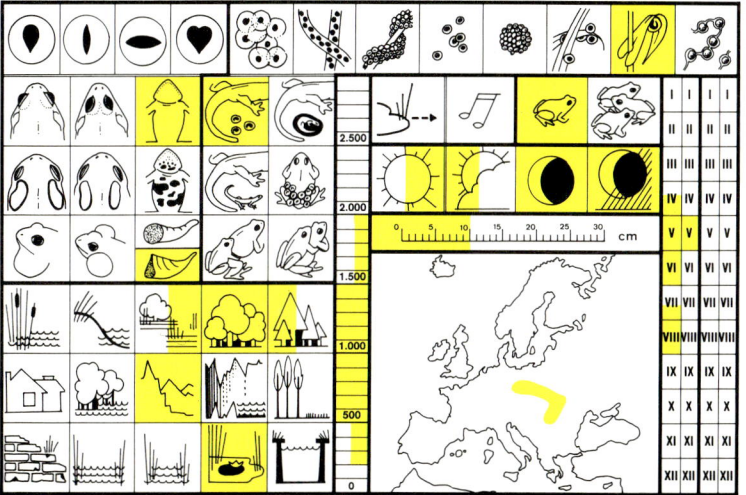

Montandon's Newt

	B	NL	L	GB	F	D	I	E	CH	A
=										
↙										
↘										
∌										
†										

Often together with *T. alpestris*. Distribution restricted to Tatras and Carpathian mountains. Seems to resist pollution to a certain extent. Female is shown bottom left.

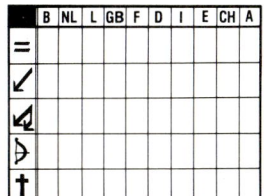

Triturus helveticus

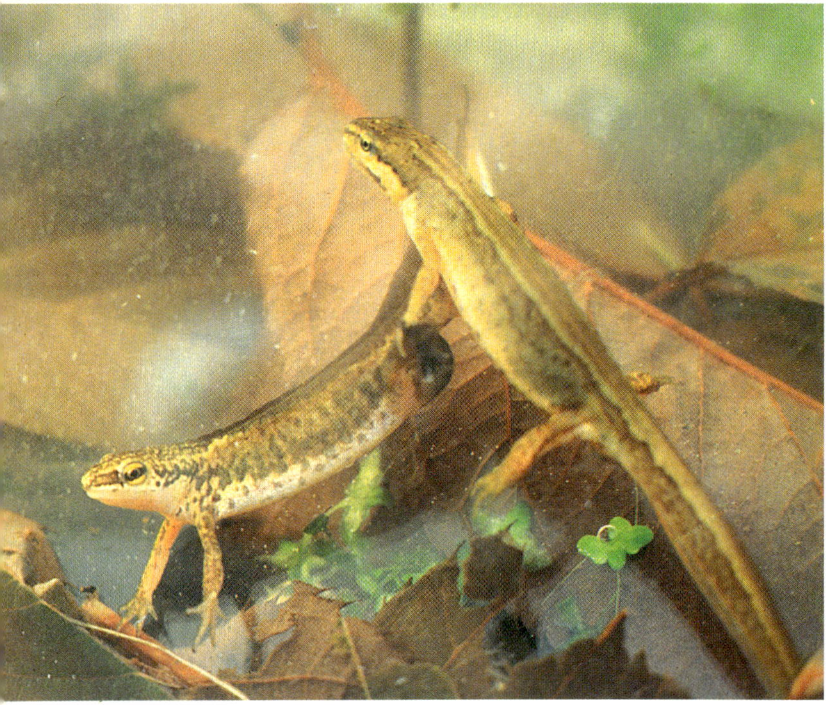

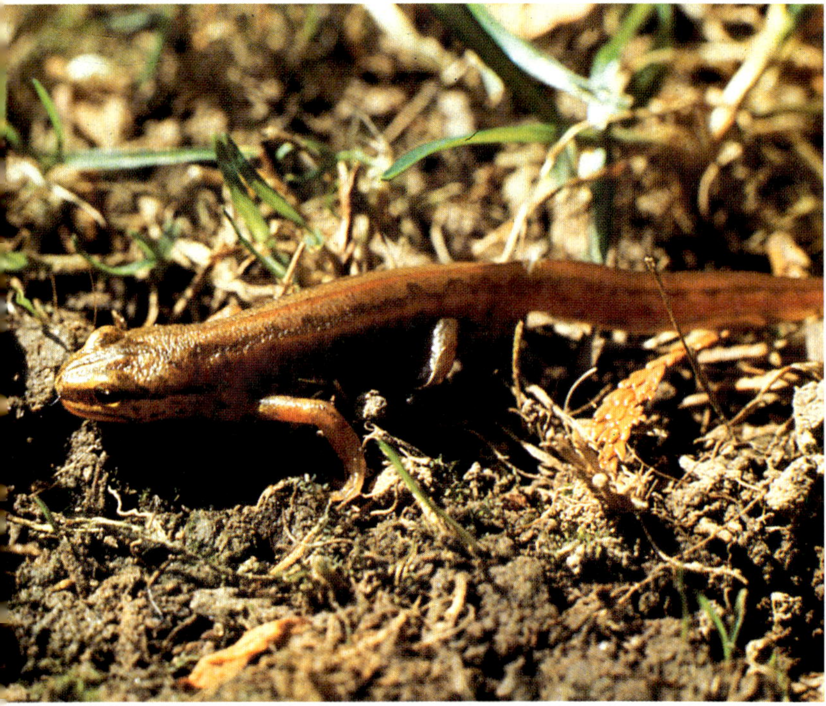

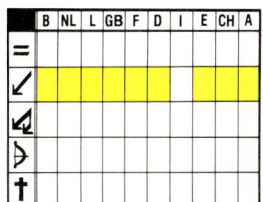

Palmate Newt

	B	NL	L	GB	F	D	I	E	CH	A
=										
↙										
◿										
⊅										
†										

Tail ends in a short filament. Male has large dark webs
on hind feet. Throat unspotted, often translucent pink.
Breeds in shallow water. Hibernates in water too.
Lays 300–400 eggs. Neoteny occurs, predominantly in
females. The terrestrial form is shown bottom left.

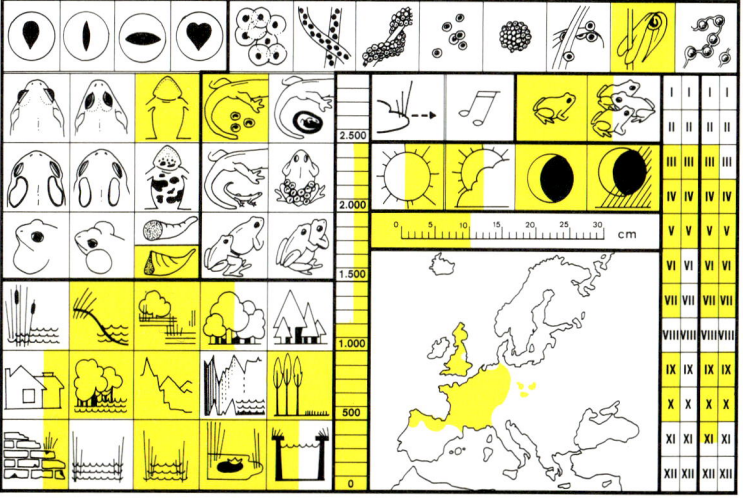

 Triturus italicus

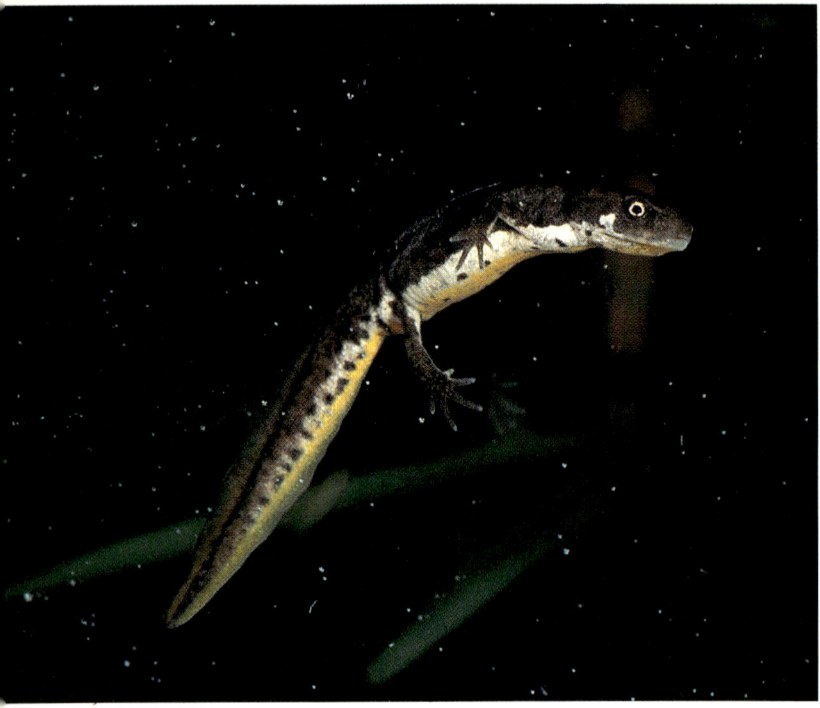

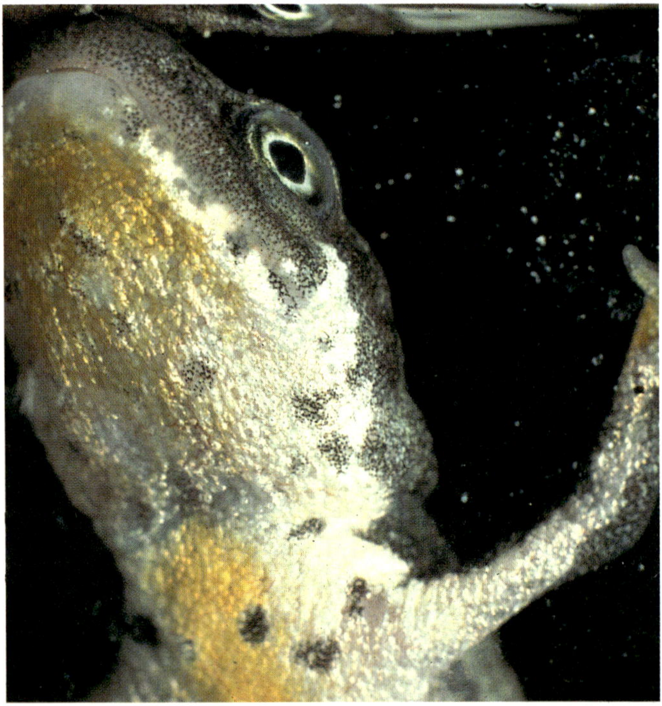

Italian Newt

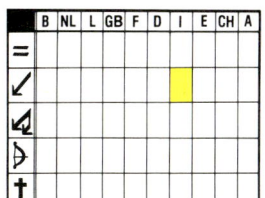

	B	NL	L	GB	F	D	I	E	CH	A
=										
⟋										
◿										
⌀										
†										

Very small animal, especially the male. Belly paler than throat (in *T. vulgaris* it is the other way round).

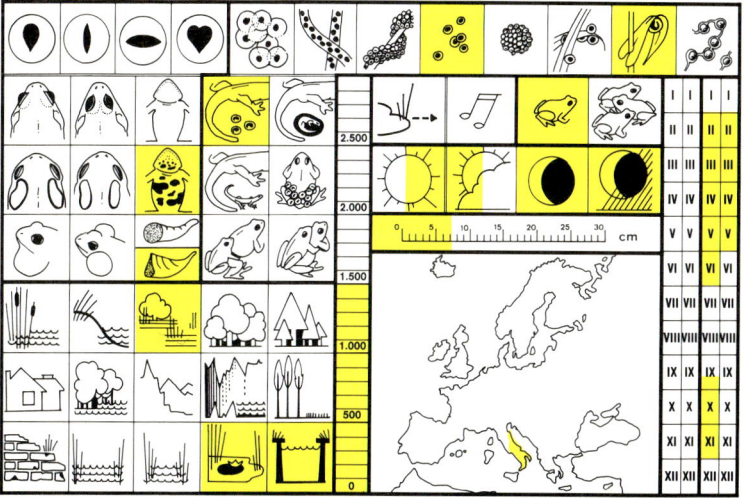

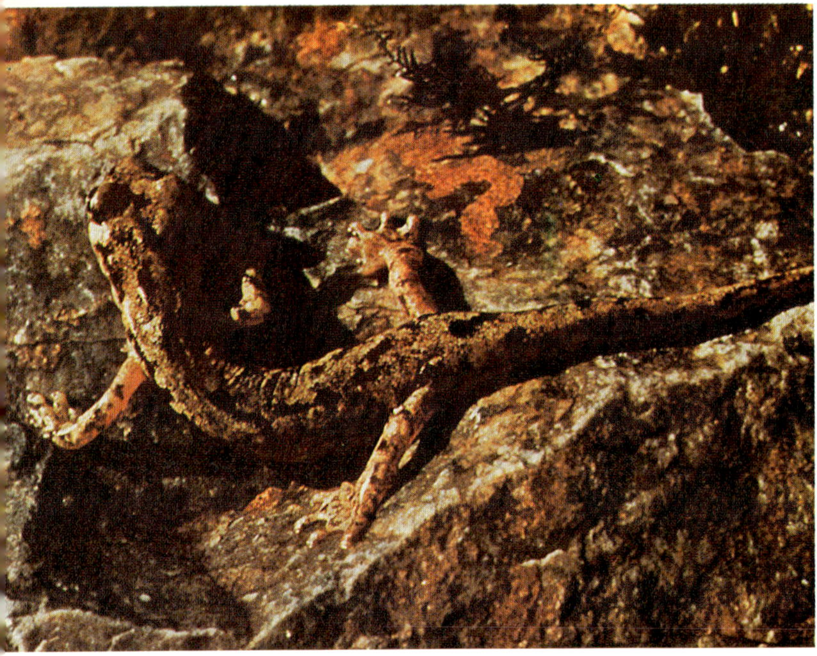

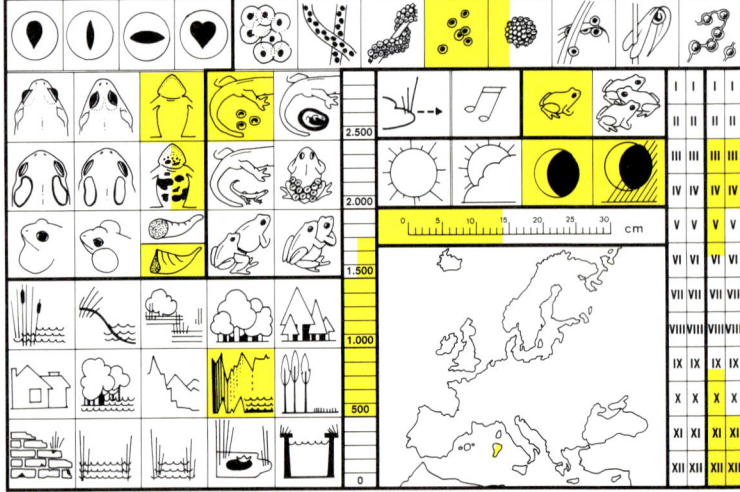

Sardinian Cave Salamander

	B	NL	L	GB	F	D	I	E	CH	A
=										
↙										
↙										
⊅										
†										

Hydromantes species have an extendable tongue to catch food. Lives in cool caves (under 17°C) with dense humidity. Climbs very well

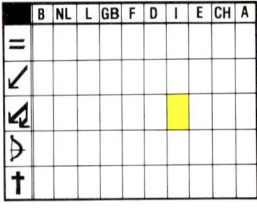

Italian Cave Salamander

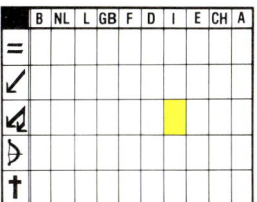

·	B	NL	L	GB	F	D	I	E	CH	A
=										
↙										
⬿						■				
↯										
†										

Smaller than preceding species. *H. italicus germani* is a subspecies of *H. italicus*. Occurs on the French border, Liguria and further southwards towards Tuscany.

21 Proteus anguinus

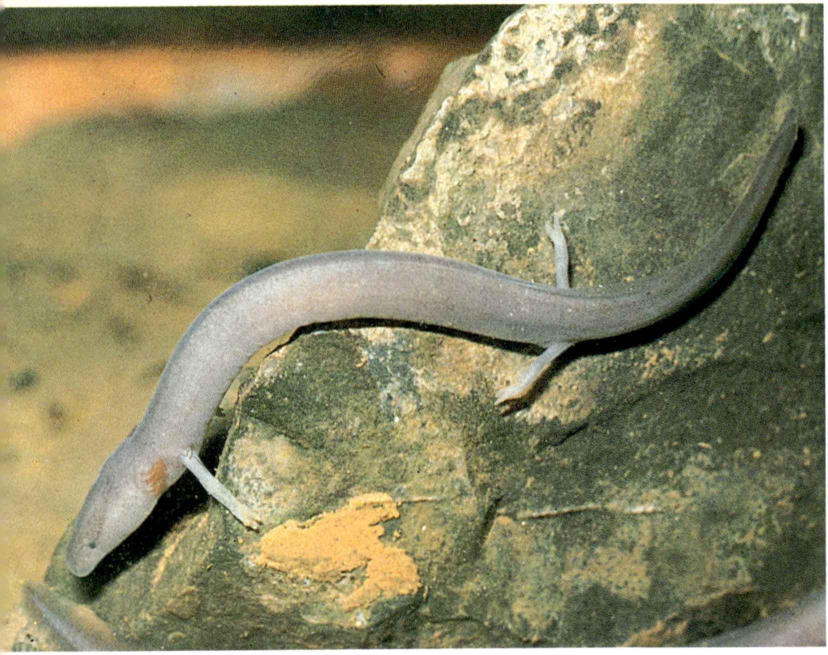

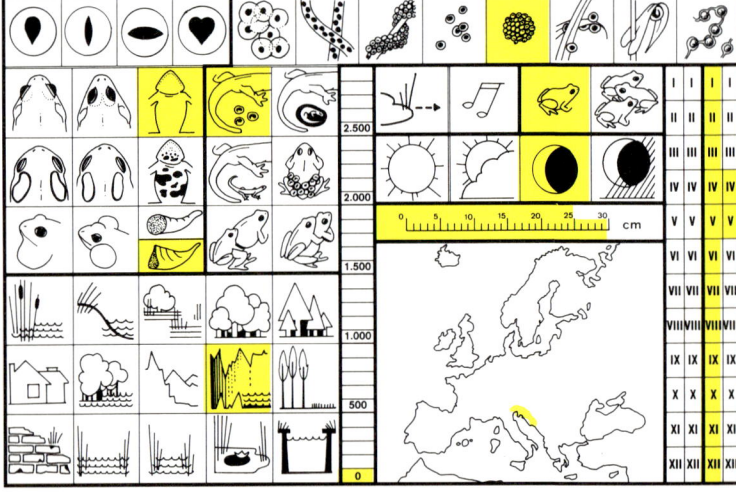

Olm

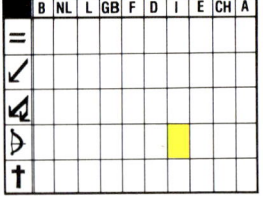

Neotenous. Gills always present. Completely aquatic species. Poorly developed limbs. Very small eyes. Young individuals faintly blotched. Lives in cold water between 5 and 10°C.

PART TWO

TOADS, PAINTED FROGS AND PARSLEY FROGS

Discoglossidae
Pelobatidae
Bufonidae

Bufo viridis

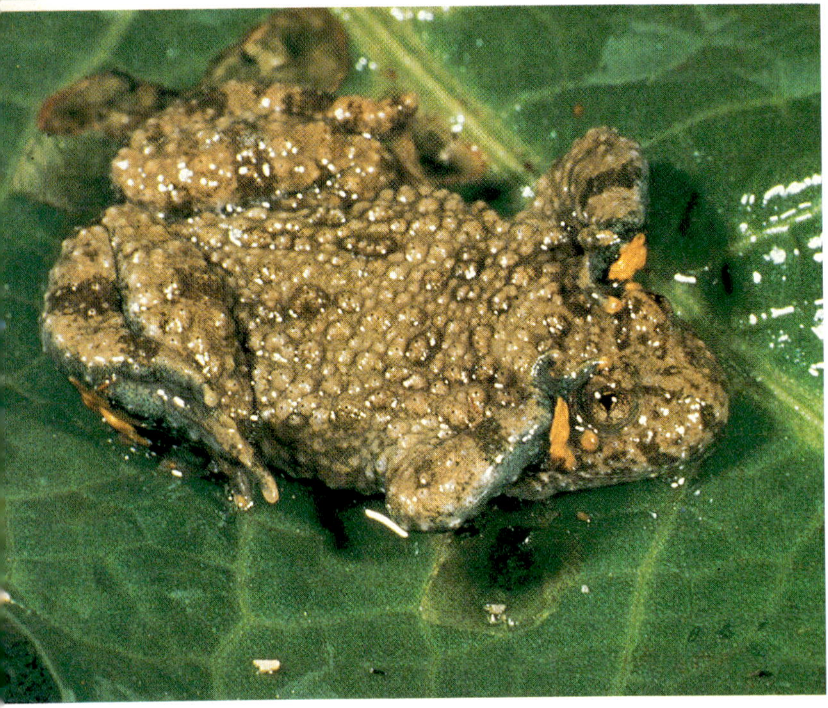

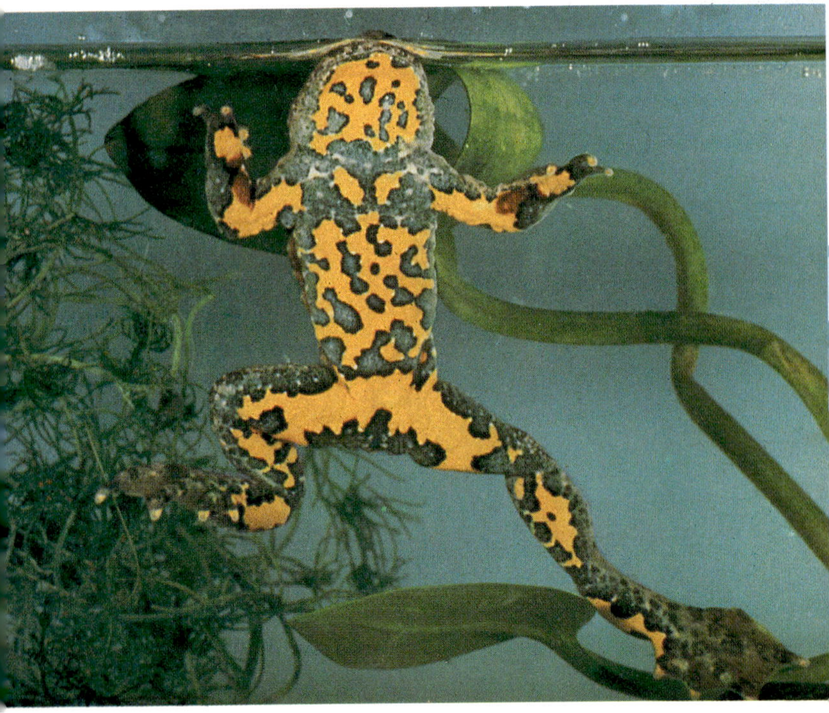

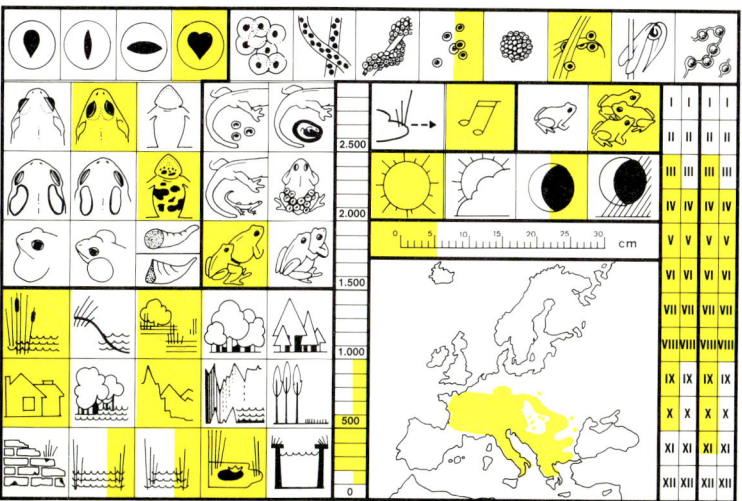

Yellow-bellied Toad

	B	NL	L	GB	F	D	I	E	CH	A
=										
↙					▣	▣				
◸				▣		▣			▣	
⟩		▣	▣							
†	▣									

Belly with yellow markings. Back warty. Typical defence behaviour. Poisonous, though no danger to man. 80–100 eggs. Larvae up to 4.5cm. Voice: a high-pitched 'oonk-oonk'.

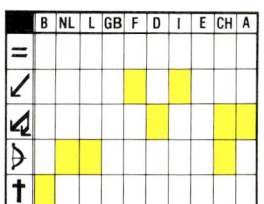

Bombina bombina

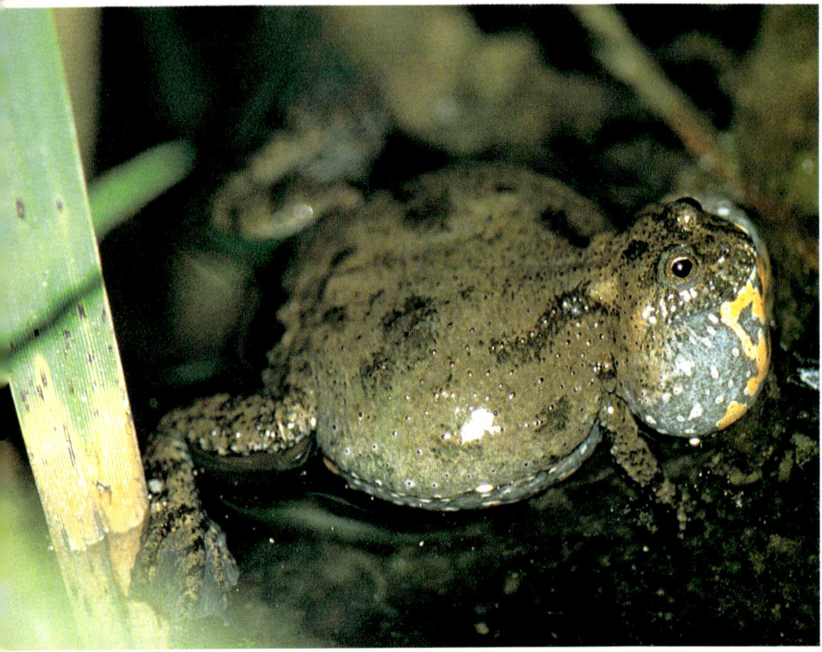

Fire-bellied Toad

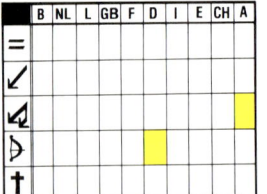

Very like preceding species. Differences are range, red colour, lower-pitched call, white dots on belly, more robust, sometimes greenish, internal vocal sacs, back *not* spiny.

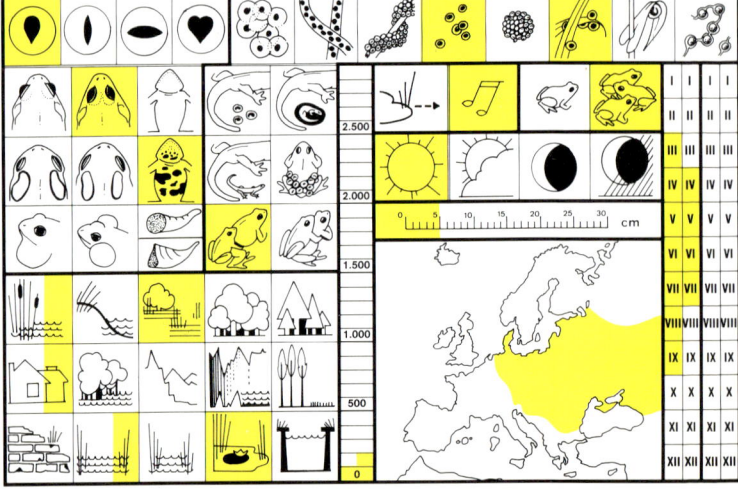

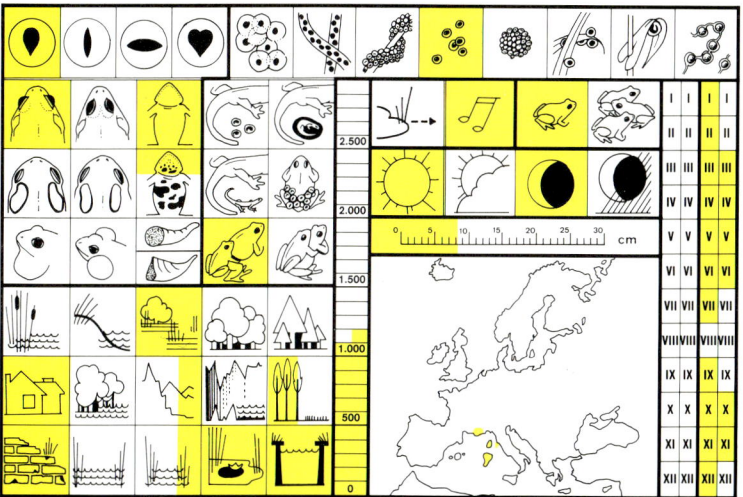

Tyrrhenian Painted Frog

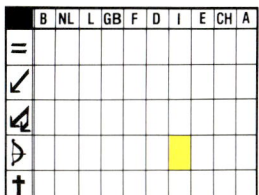

	B	NL	L	GB	F	D	I	E	CH	A
=										
↙										
◿										
⸮										
†										

Very similar to *D. pictus*. Lays between 300 and 1,000 eggs.

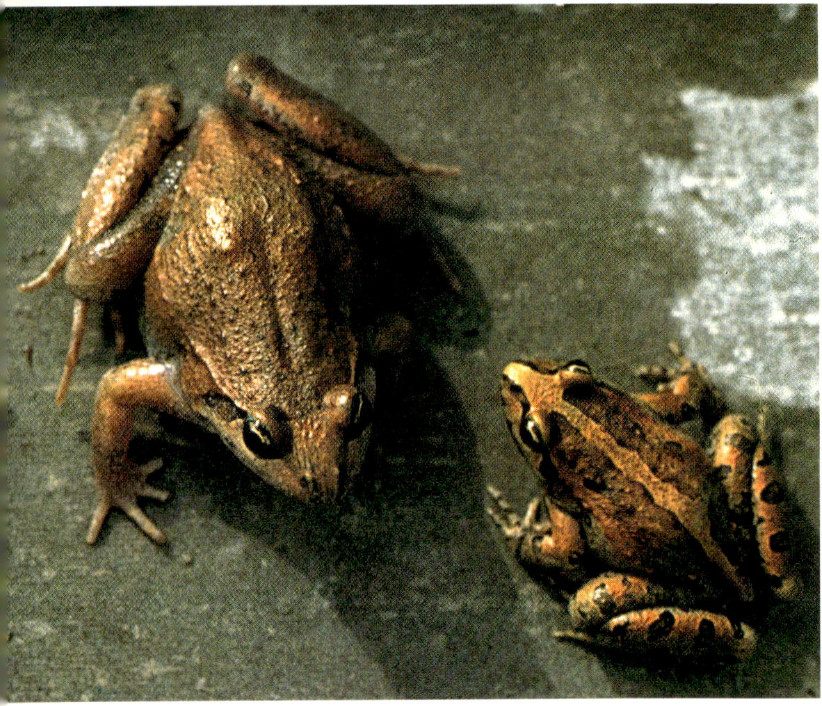

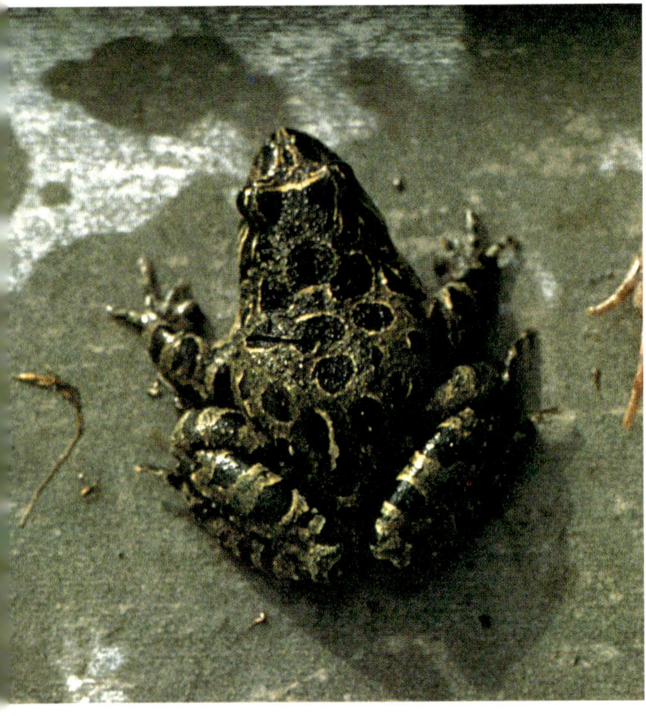

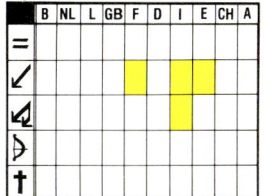

Painted Frog

	B	NL	L	GB	F	D	I	E	CH	A
=										
↙					▓		▓			
◩						▓				
⋺										
†										

Often aquatic way of life. Can be encountered even in rainwater cisterns. Lays nearly 1,000 eggs, which hatch in 2–8 days. Metamorphosis in one month. 'Pictus' and 'striatus' forms are shown top left, with 'ocellatus' form bottom left.

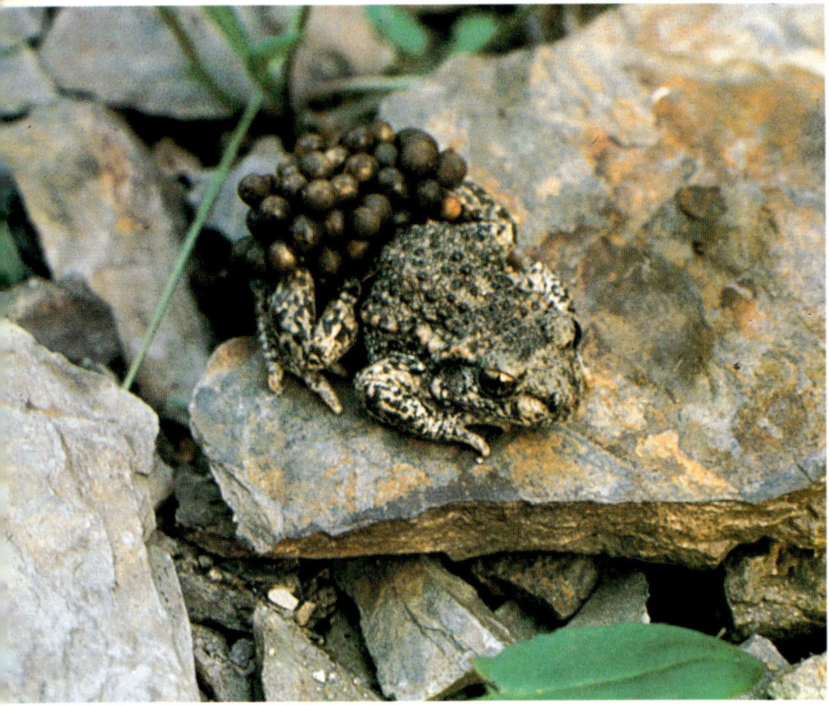

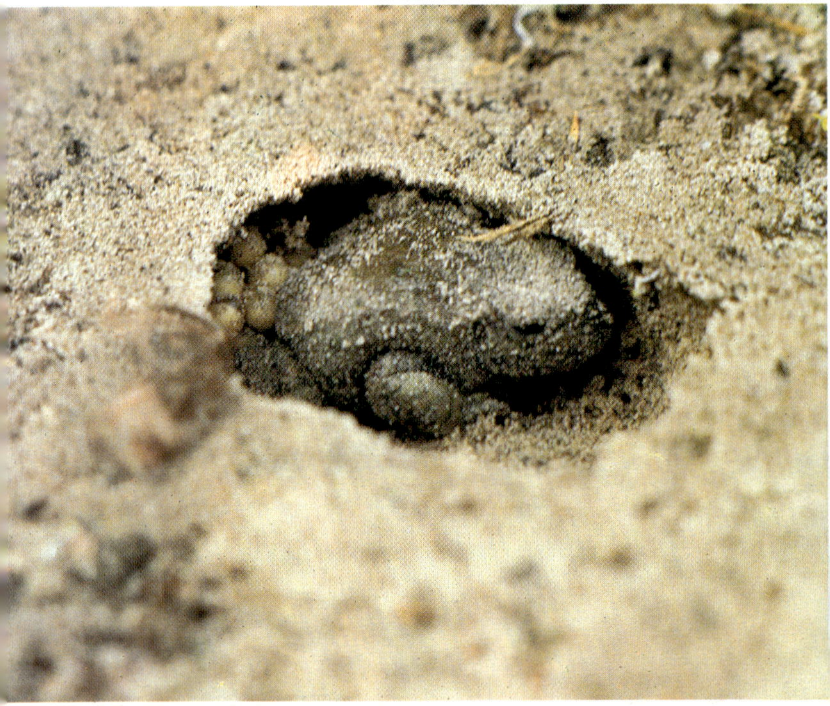

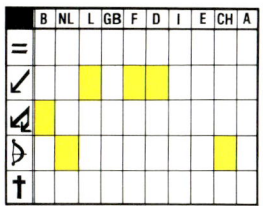

Midwife Toad

	B	NL	L	GB	F	D	I	E	CH	A
=										
↙			▓	▓	▓					
◿	▓									
▷	▓								▓	
†										

Very secretive, often in burrows under logs or roots. Typical call resembles a tiny whistle. Larvae often hibernate. Male carries eggs.

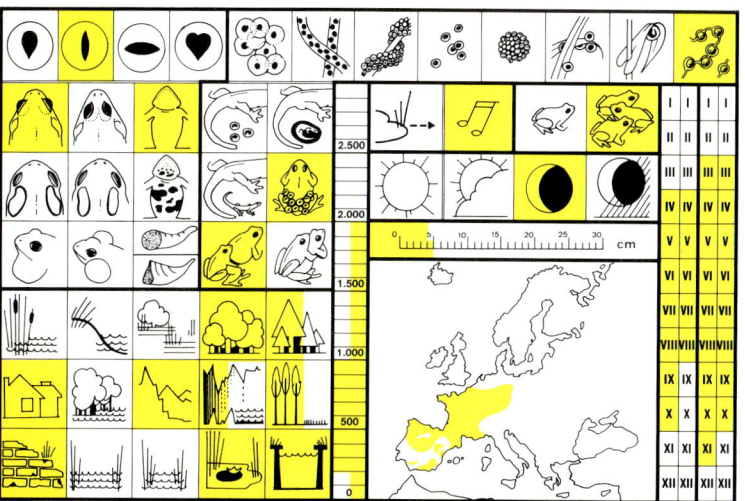

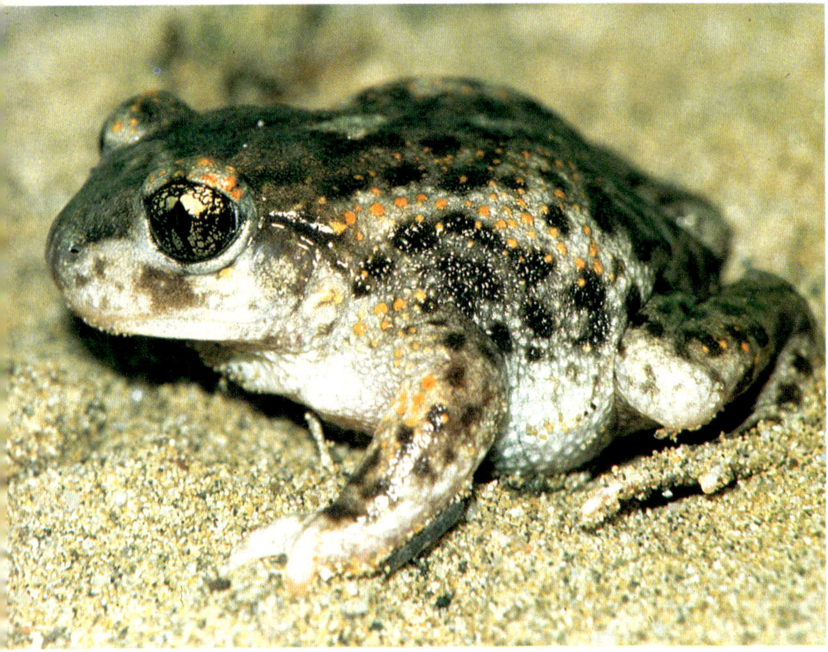

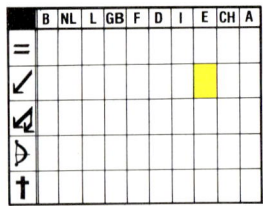

Iberian Midwife Toad

	B	NL	L	GB	F	D	I	E	CH	A
=										
↙								▨		
◹										
▷										
†										

Very similar to *A. obstetricans*. Difference can be seen in the underside of the hand.

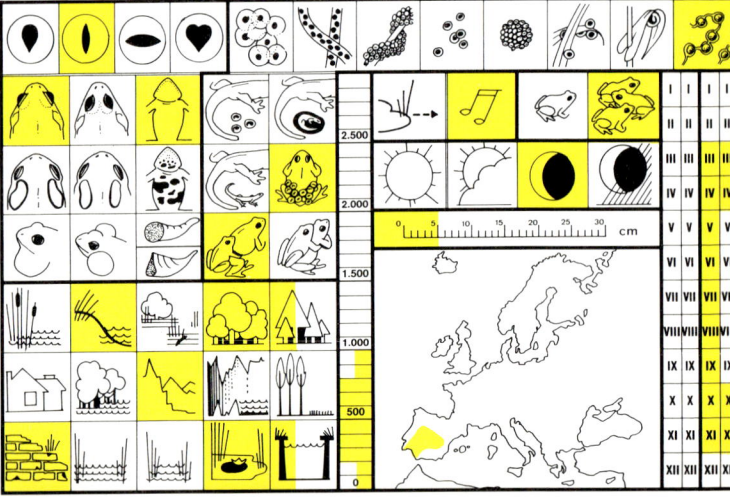

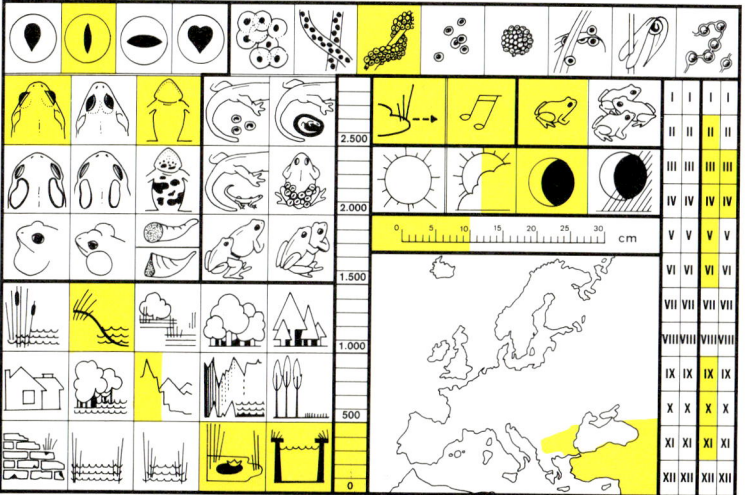

Eastern Spadefoot

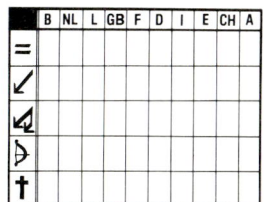

	B	NL	L	GB	F	D	I	E	CH	A
=										
✓										
◁										
▷										
†										

Habitat and habits similar to other Spadefoots. Eyes more protruding than in other Spadefoots. No webs between toes of hind foot.

29 Pelobates fuscus

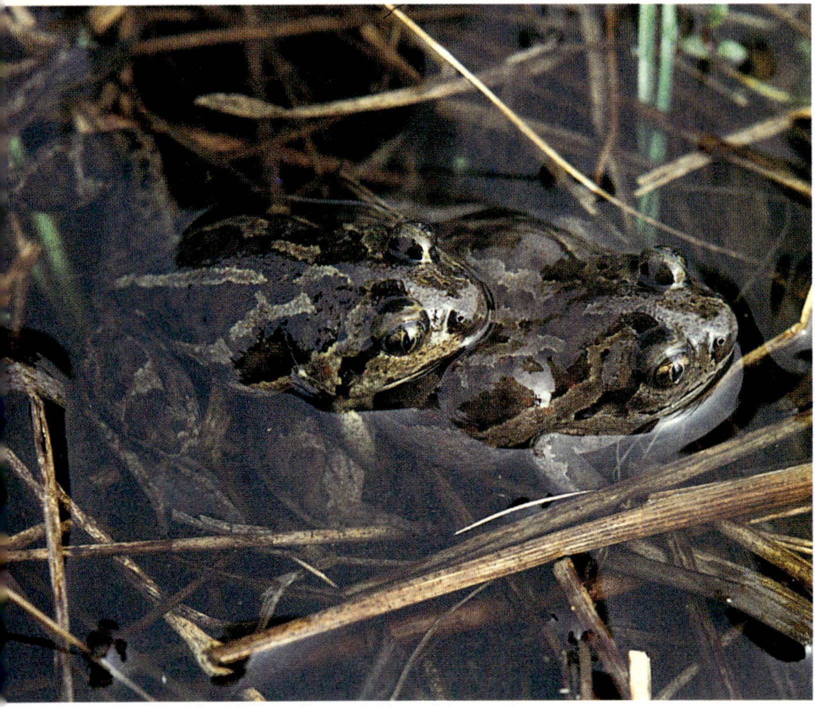

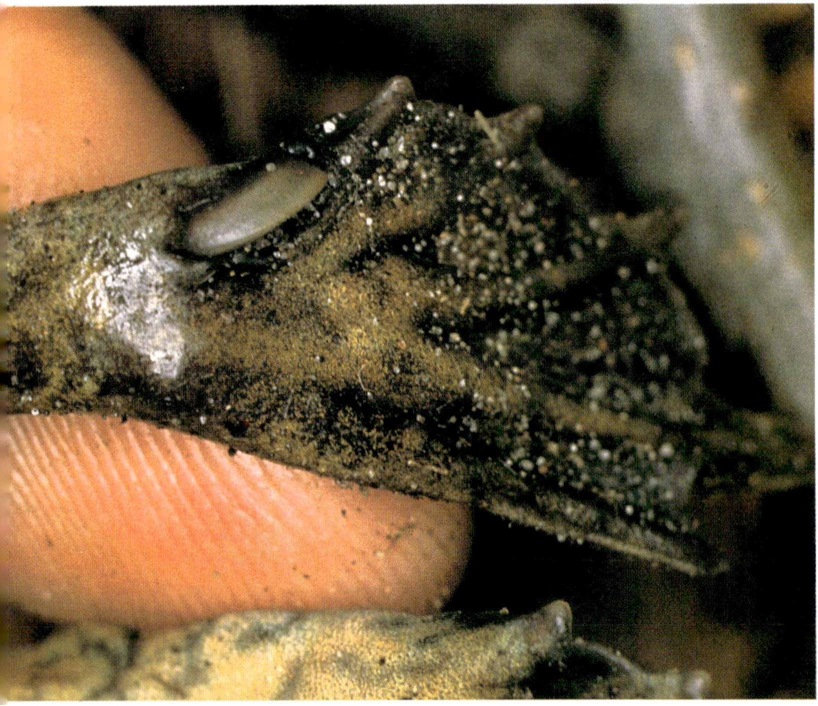

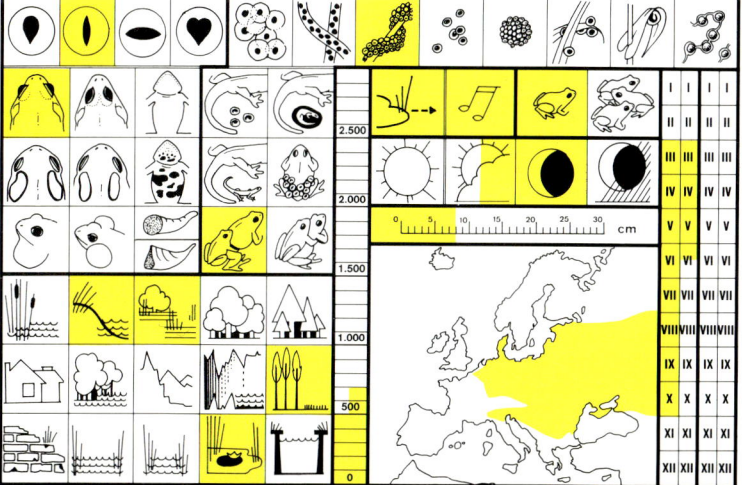

Common Spadefoot

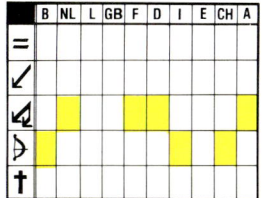

	B	NL	L	GB	F	D	I	E	CH	A
=										
↙										
◁										
♭										
†										

Frequently burrowing. Metatarsal tubercle used for digging. Lives in sandy areas. Defence secretion smells of garlic. 700–1,000 eggs, strings 40 to 70cm. Eggs often fatally infected by mildew, especially in oligotrophic waters. Voice: a repeated 'c'lock-c'lock'. Also has a shrill alarm cry. Jumps at intruders. Female is shown above (left) with male (right).

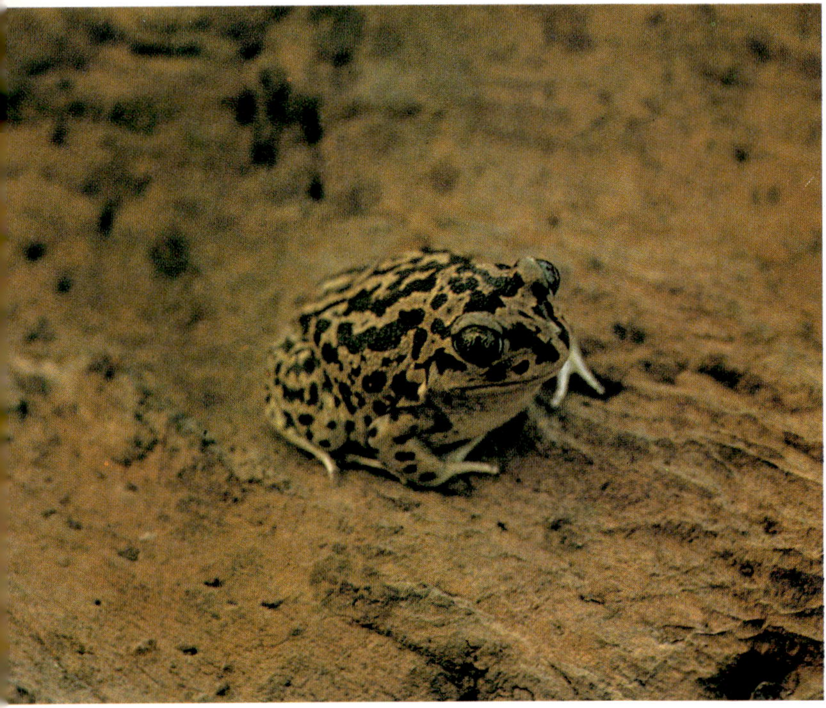

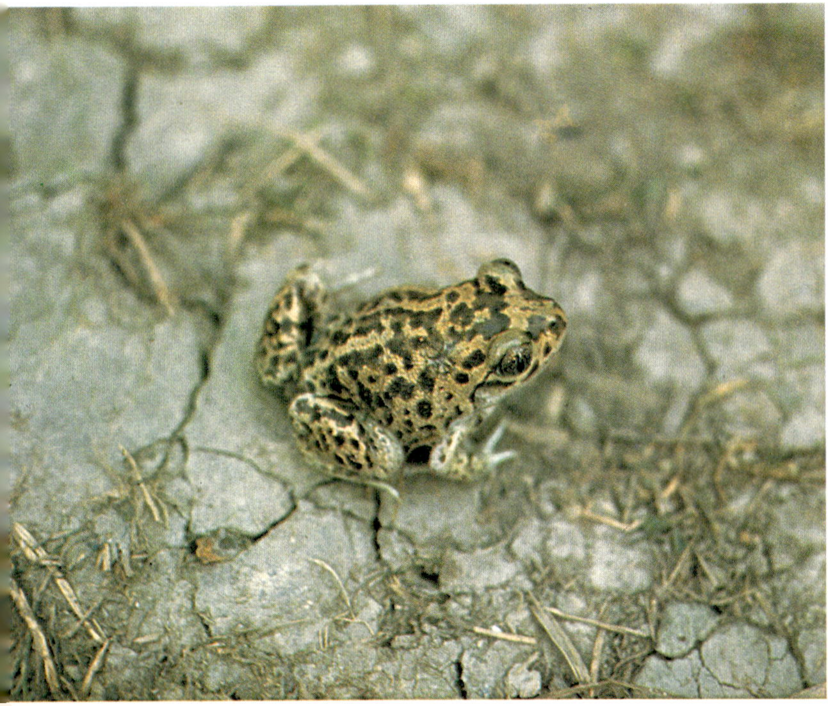

Western Spadefoot

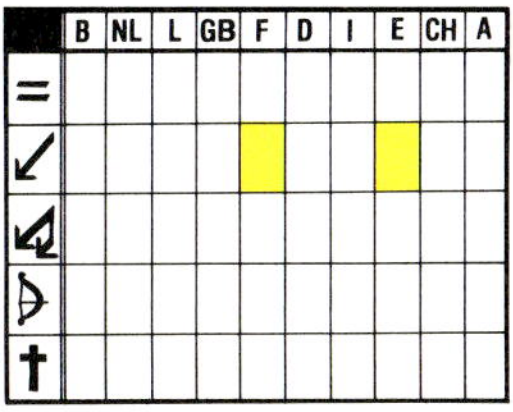

	B	NL	L	GB	F	D	I	E	CH	A
=										
↙					▨			▨		
◺										
◗										
✝										

Male smaller (7.5cm). Tadpoles of all *Pelobates* species up to 15cm epecially hibernating larvae. Numbers fluctuate year by year, according to rainfall. Call like that of a clucking hen.

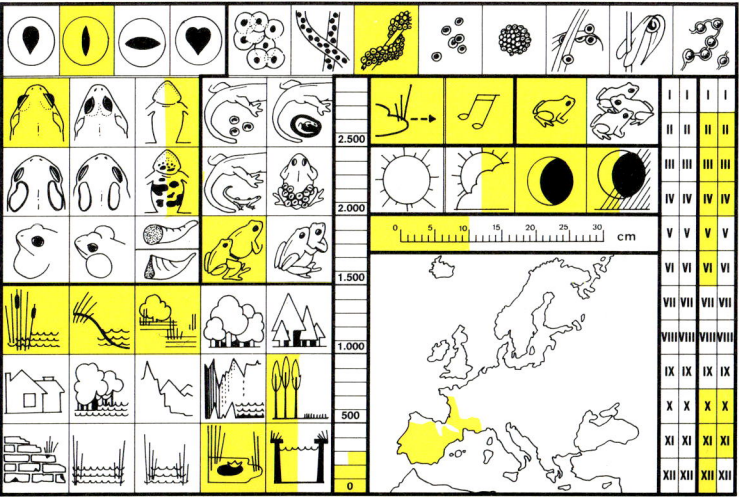

 Bufo bufo bufo

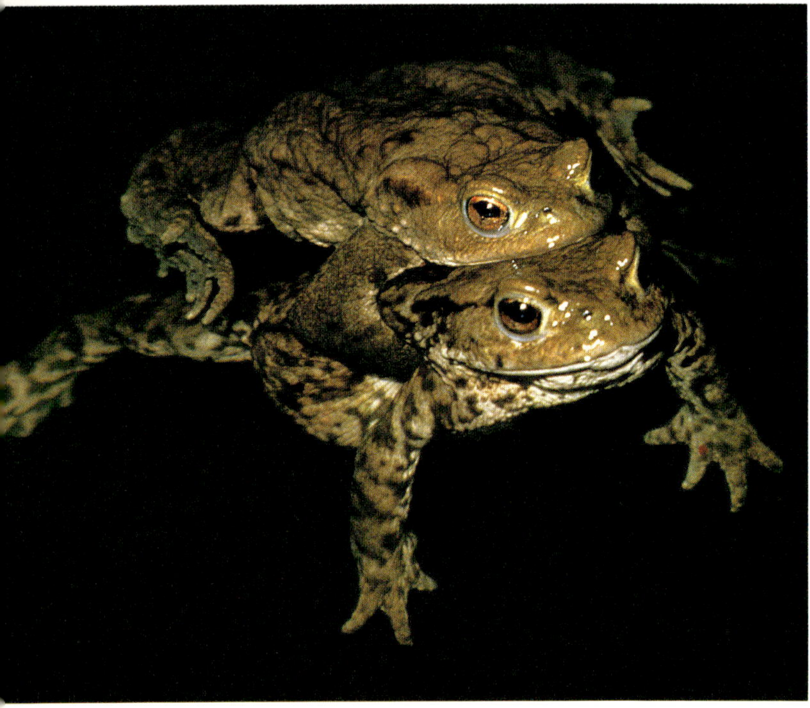

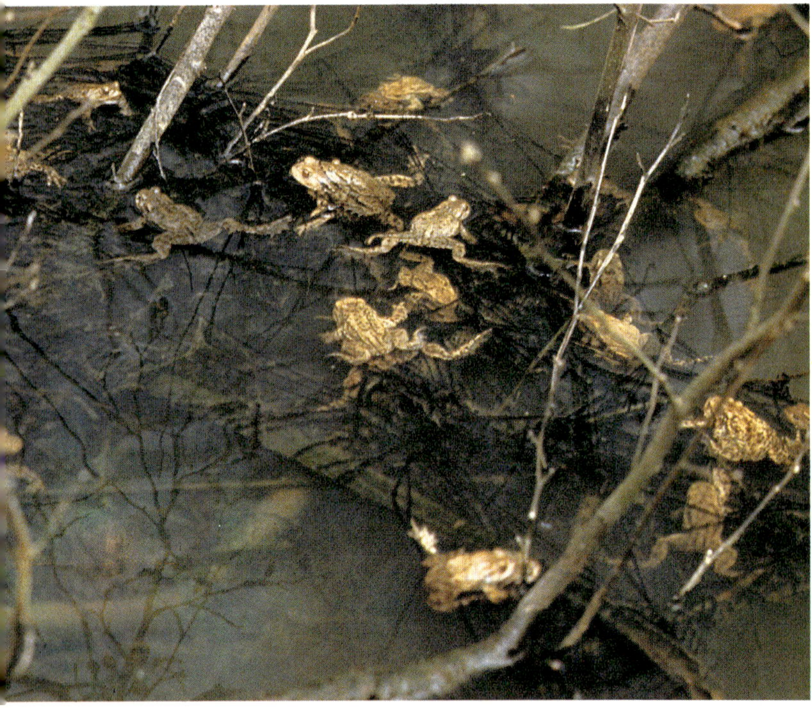

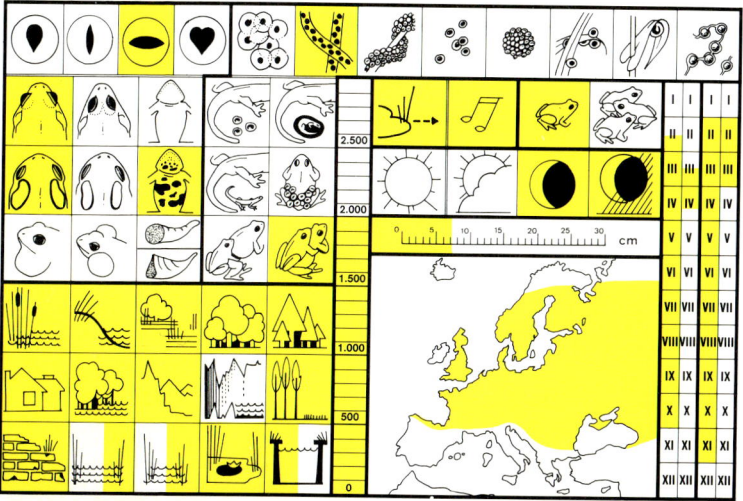

Common Toad

	B	NL	L	GB	F	D	I	E	CH	A
=										
↙										
◁										
◁										
†										

Widely distributed in temperate zones of the British Isles, western Europe and as far as Japan. Up to 7,000 eggs in strings of 3–5m. Young hatch after 2–3 weeks, metamorphose after 3–4 months. Many road victims in spring. Frequent infections from flies (*Lucilla bufonivora*) in hot summers in north-west Europe.

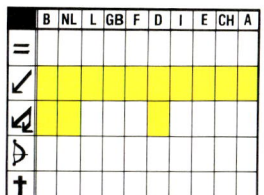

32 Bufo bufo spinozus

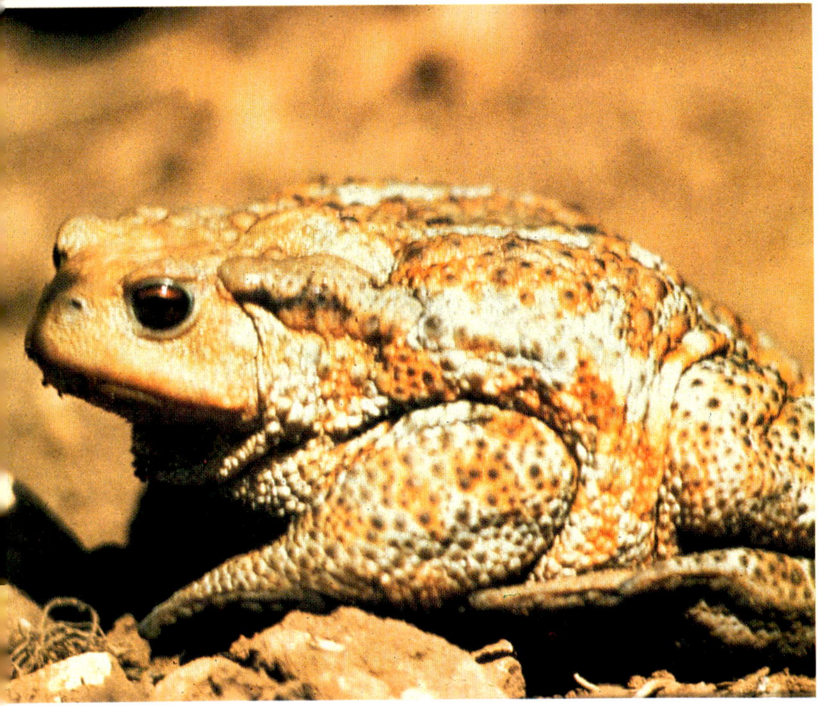

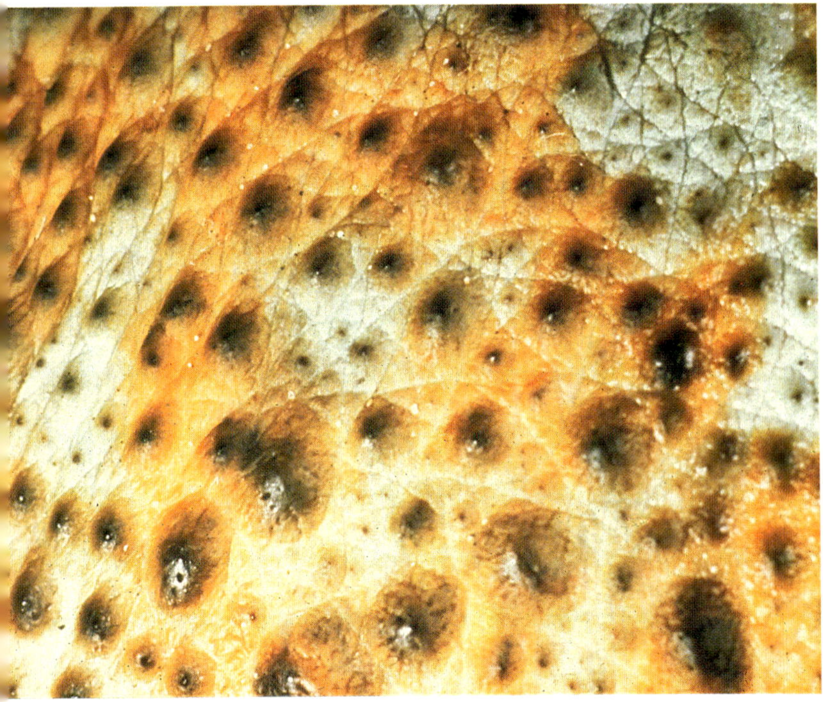

Southern Common Toad

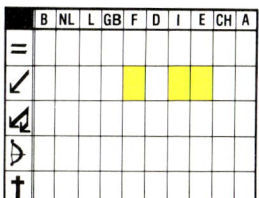

Southern subspecies of *Bufo bufo*. Much larger and has round instead of cone-shaped warts on the skin. Follows human habitations. In summer in moist places (farms, irrigation systems, camp-site showers). Also in dry habitats. Female is shown top left.

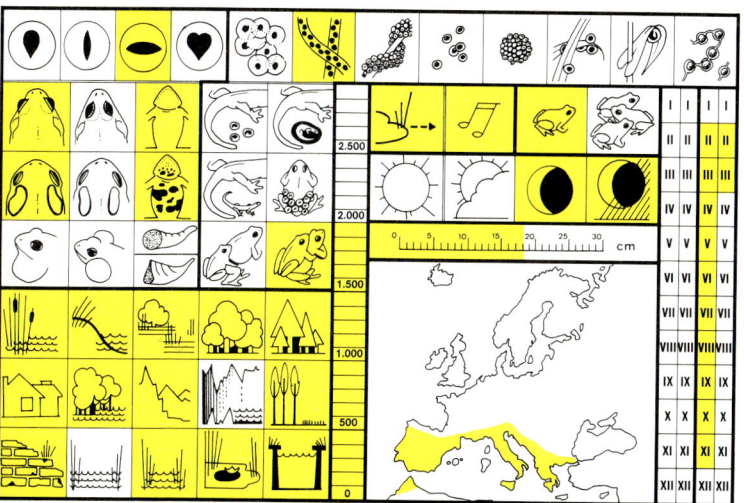

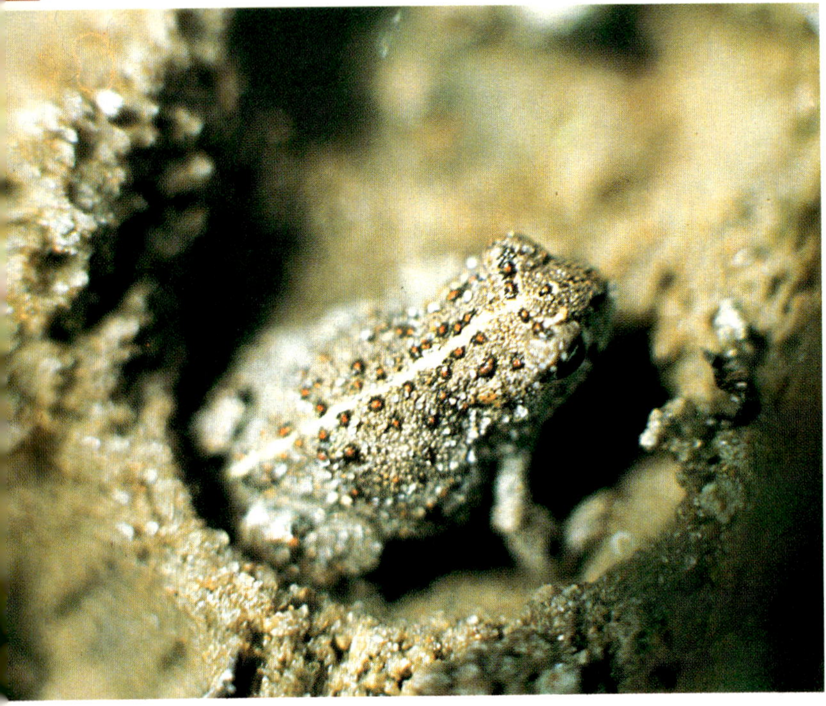

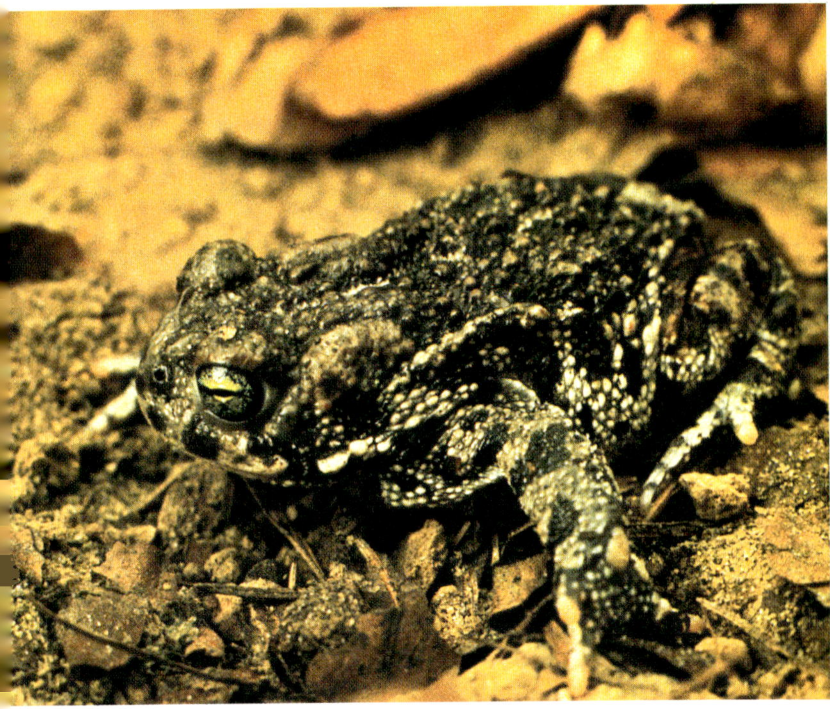

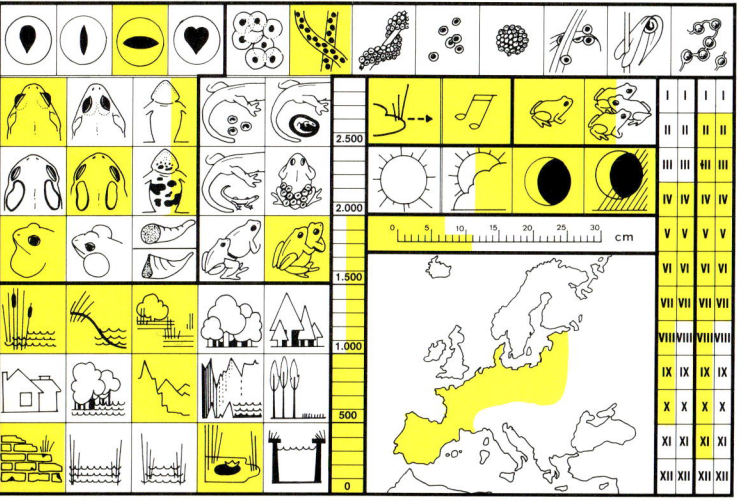

Natterjack

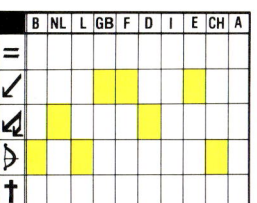

	B	NL	L	GB	F	D	I	E	CH	A
=										
↙				█		█				
◺		█			█					
▷		█						█		
†										

Burrows frequently. Moves by running rather than hopping. Resists heat and drought well. Pioneer of newly-created sandy soils. Can breed even in brackish water. 3,000–4,000 eggs in strings. These hatch after 4–6 days. Metamorphosis in July–August. Young often diurnal to prevent cannibalism by nocturnal adults. Whole populations may suddenly migrate. Juvenile is shown top left, with the winter form bottom left.

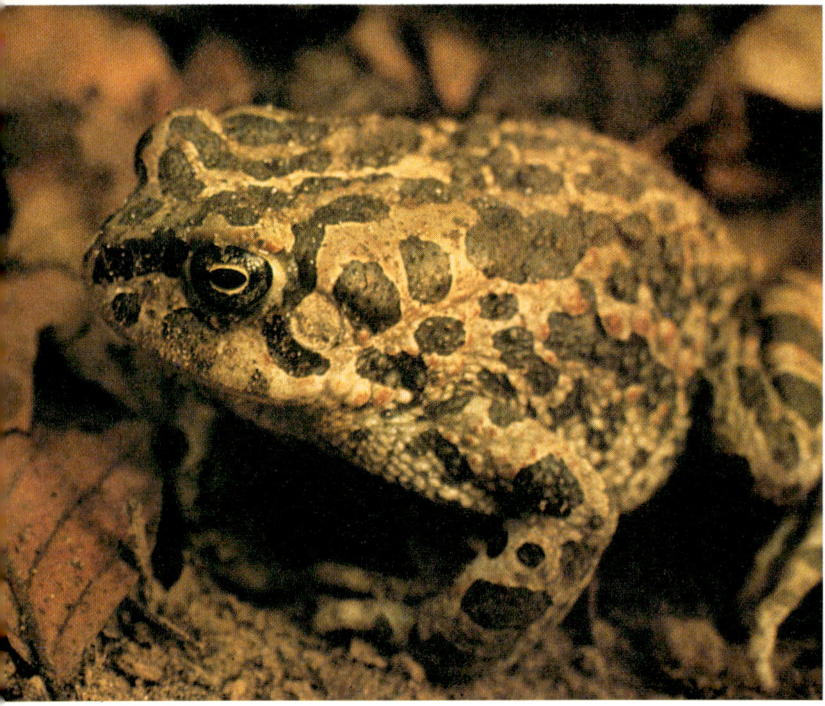

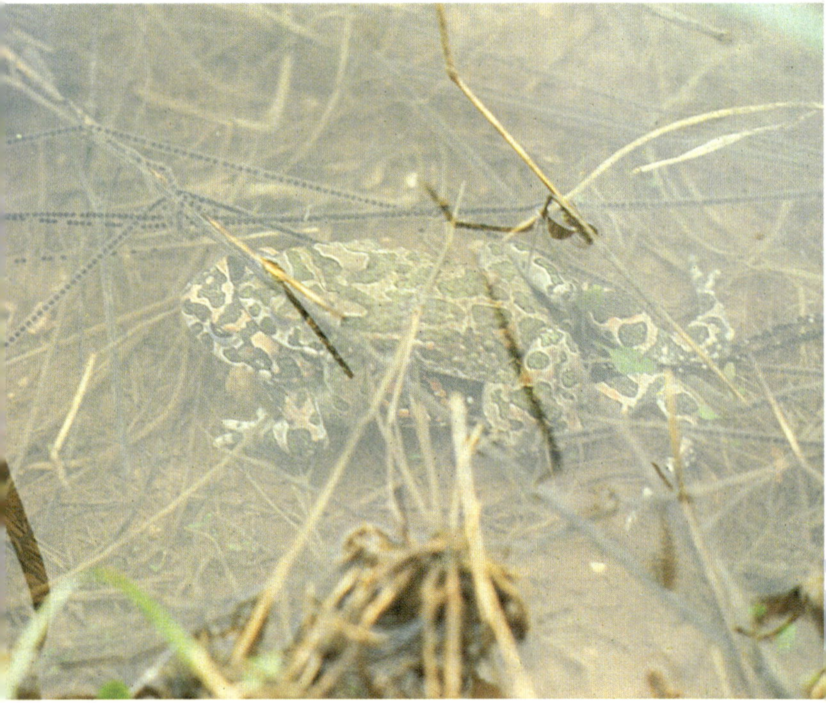

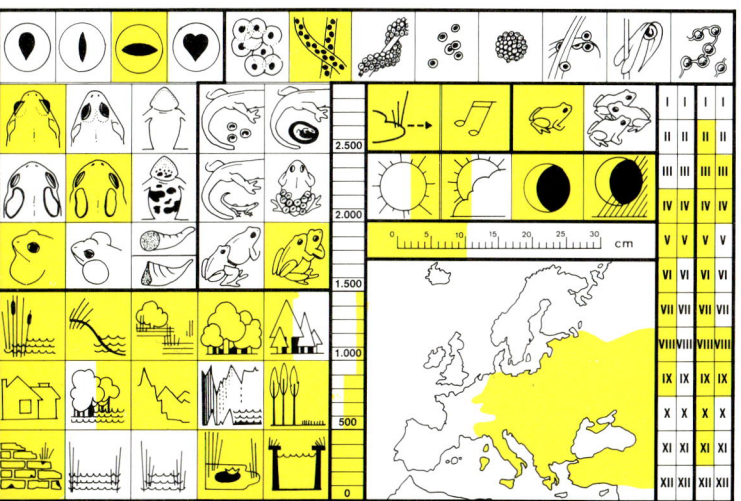

Green Toad

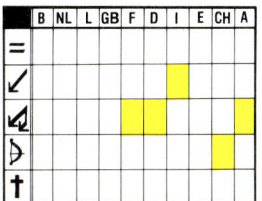

	B	NL	L	GB	F	D	I	E	CH	A
=										
↙							▨			
◤						▨	▨		▨	
▷								▨		
†										

Green iris, clearly visible eardrum. Follows human habitations. Is sometimes seen catching insects near street lamps in villages. Can breed in organically polluted water. Very fertile: 10,000–12,000 eggs, which hatch after 4–5 days. Metamorphosis after 2–3 months. Female is shown top left and above (lower left).

![Parsley Frog photograph]

Parsley Frog

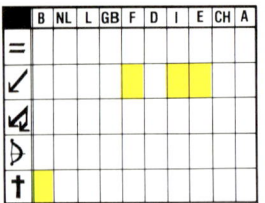

Also lives among bushes, climbs well. But no adhesive pads as in Tree Frogs. 1,000–1,600 eggs. Tadpole up to 6.5cm. Typical call: 'co-ak, co-ak' in water and 'creck-creck' on land.

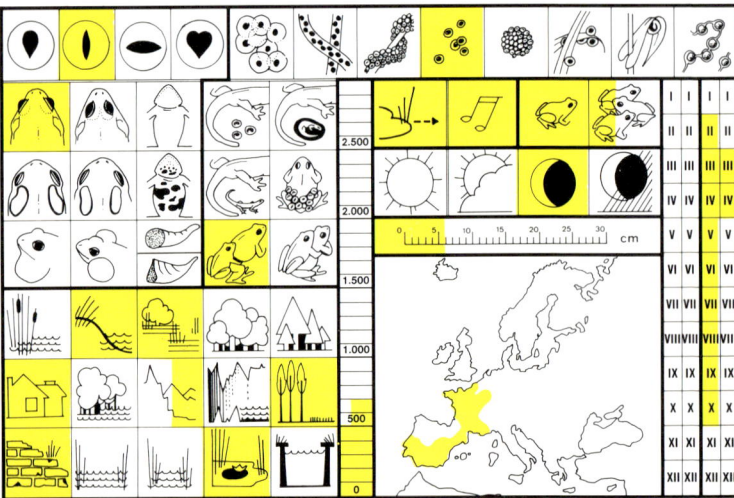

PART THREE

FROGS
AND TREE FROGS

Hylidae
Ranidae

Hyla arborea

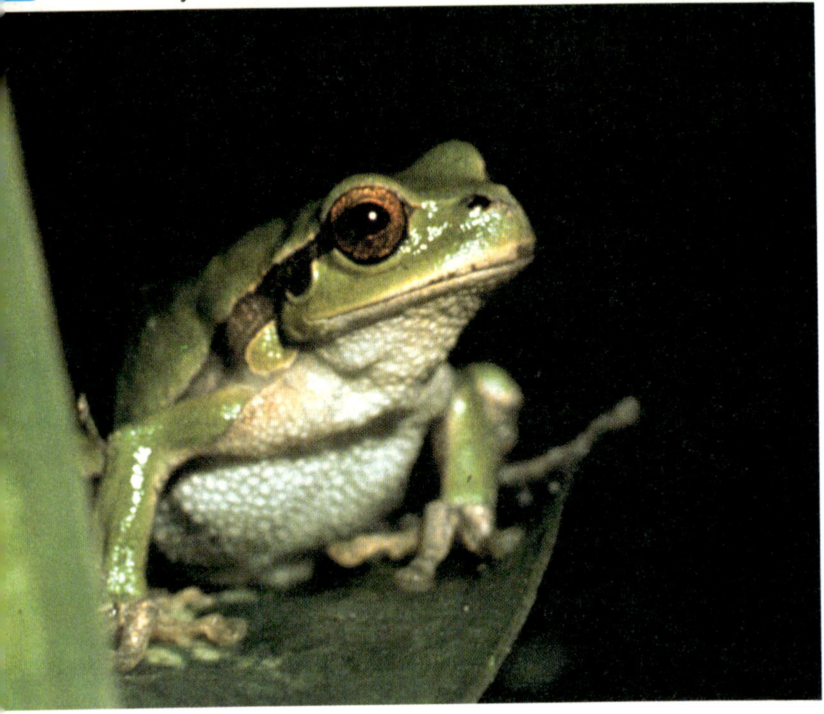

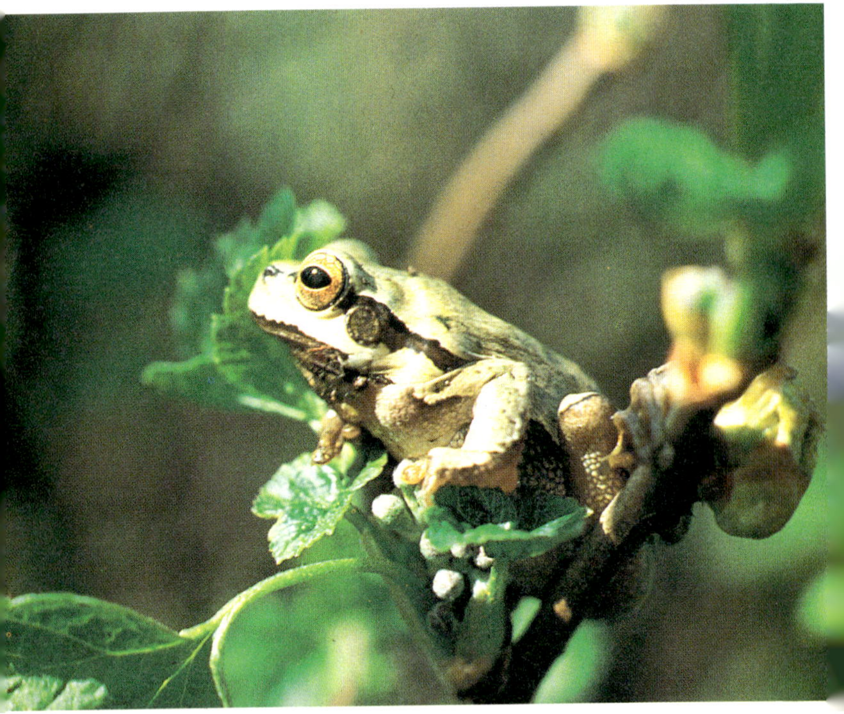

Common Tree Frog

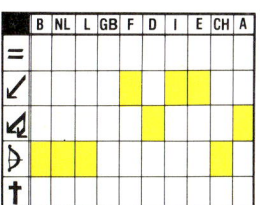

	B	NL	L	GB	F	D	I	E	CH	A
=										
↙					▓		▓	▓		
◁						▓				
▷	▓	▓	▓					▓		
†										

Perfect camouflage. Climbing species, in bushes, high grass, reeds. Adhesive pads on toes. In summer close to water. Lays 800–1,000 eggs. Noisiest species in Europe: a loud 'keck-keck-keck' in choruses. Some individuals lack yellow pigment and are blue. Female is shown top left.

37 Hyla meridionalis

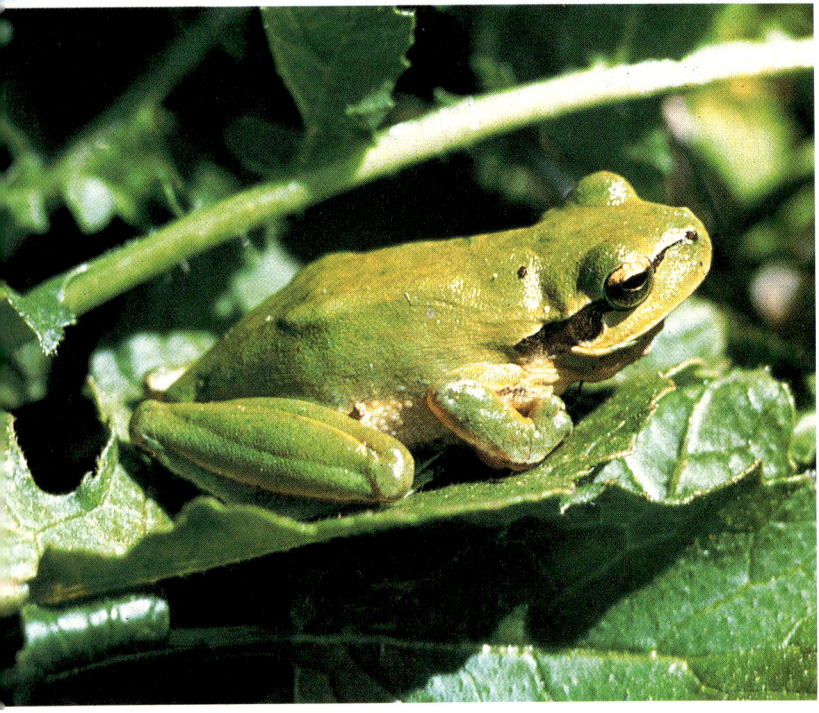

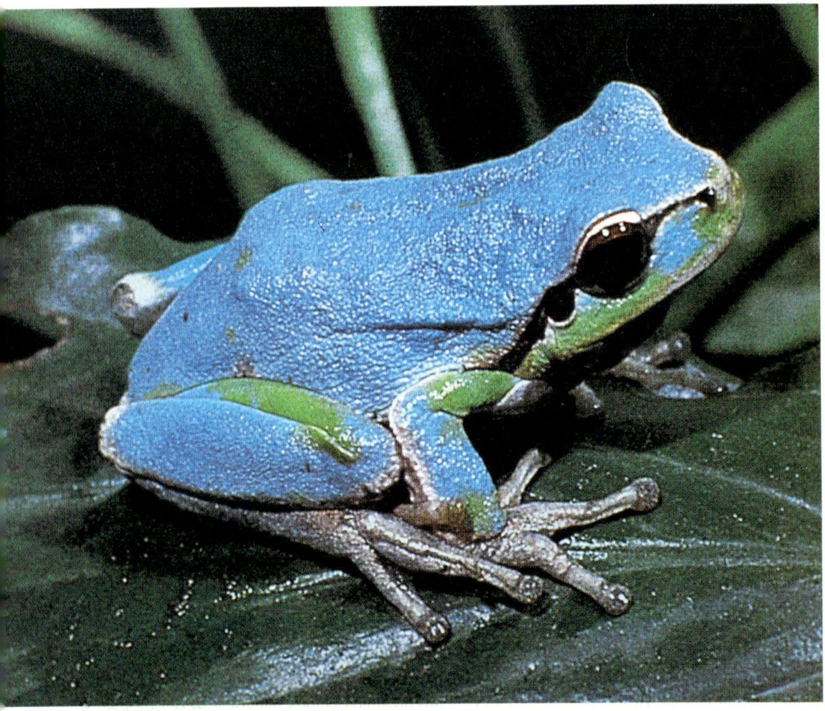

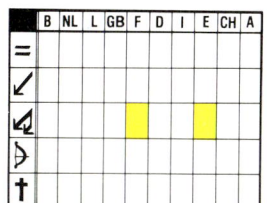

Stripeless Tree Frog

	B	NL	L	GB	F	D	I	E	CH	A
=										
↙										
◁										
ᕗ										
†										

Similar to *H. arborea* but without black stripe on flanks. Calls faster than *H. arborea*. Frequently captured for pet trade, although protected. The blue form is shown bottom left.

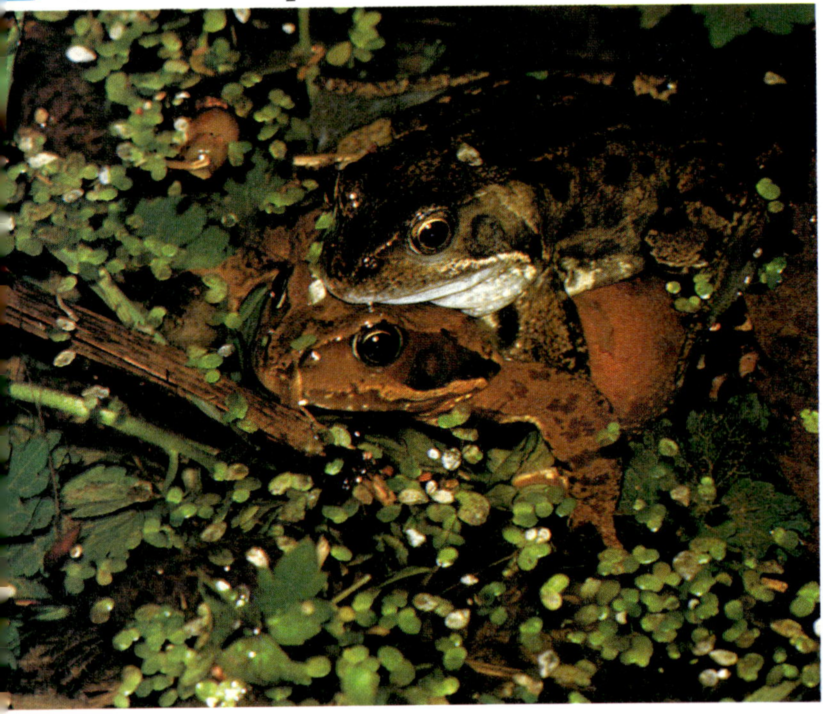

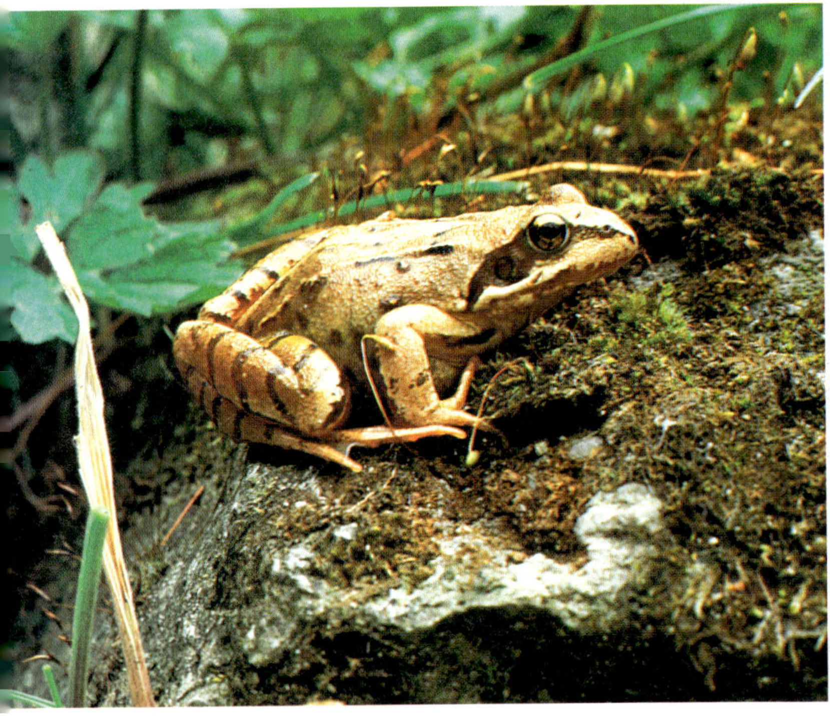

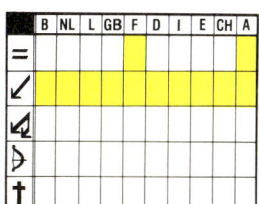

Common or Grass Frog

	B	NL	L	GB	F	D	I	E	CH	A
=										
╱										
◿										
ϸ										
†										

Largest Brown Frog. Widely distributed. 1,000–4,000 eggs per brood, depending on body size. Captured in large numbers for laboratories and for frog's legs. Female is shown bottom left.

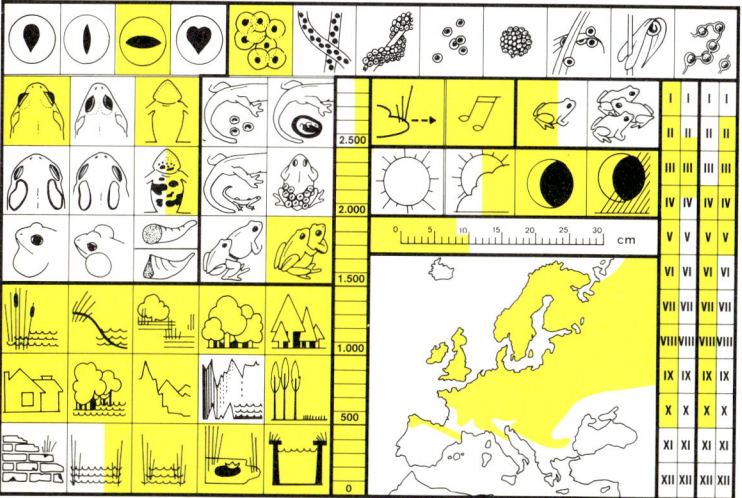

39 Rana dalmatina

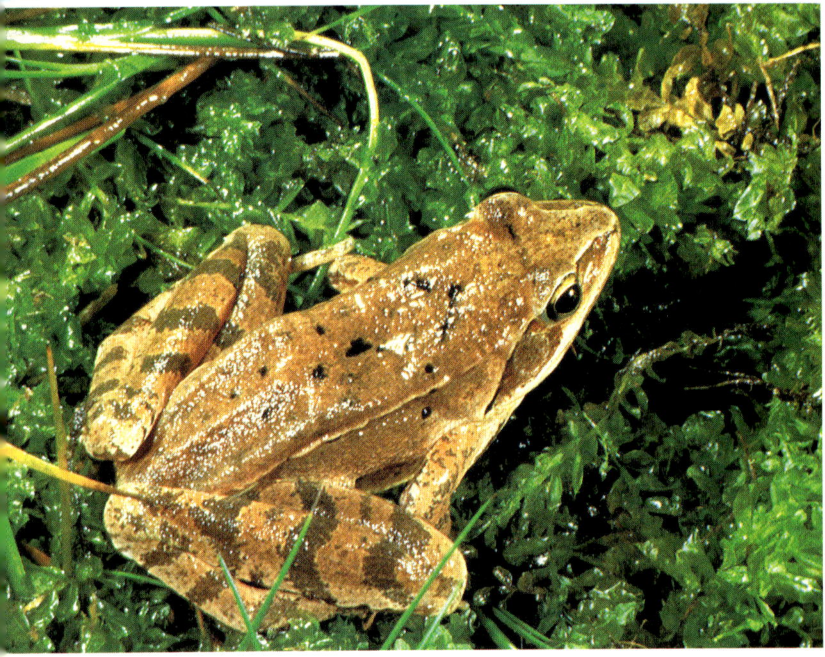

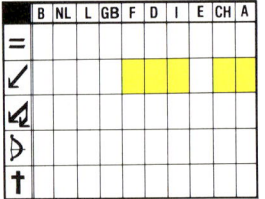

Agile Frog

Very fast. Very similar to *R. temporaria* but with larger metatarsal tubercle and large eardrum close to eye 1,000–2,000 eggs.

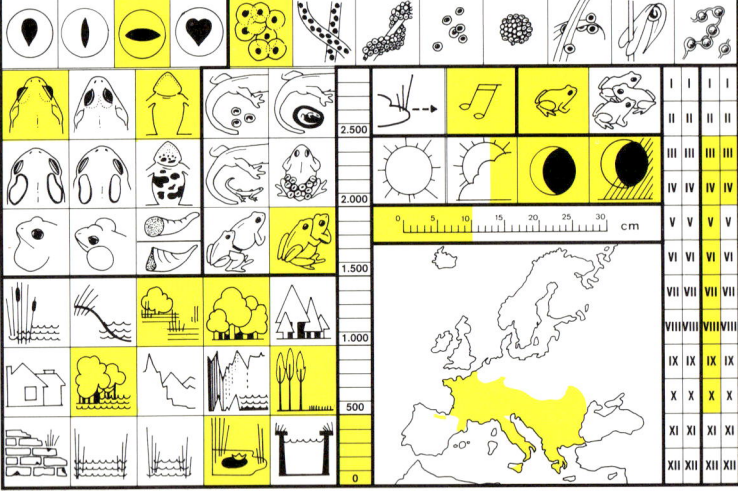

Moor Frog

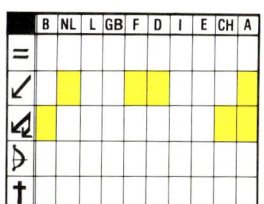

	B	NL	L	GB	F	D	I	E	CH	A
=										
↙		▨			▨	▨				▨
◣	▨								▨	
▷										
†										

Smaller than other Brown Frogs. Broad pale vertebral stripe. Male blue in mating season. 600–1,200 eggs. Voice: as of air escaping from a submerged bottle. Sexually mature after 3 years.

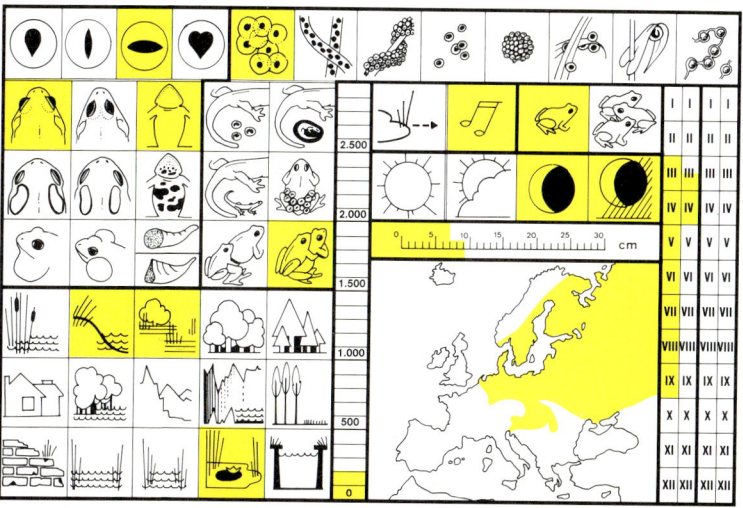

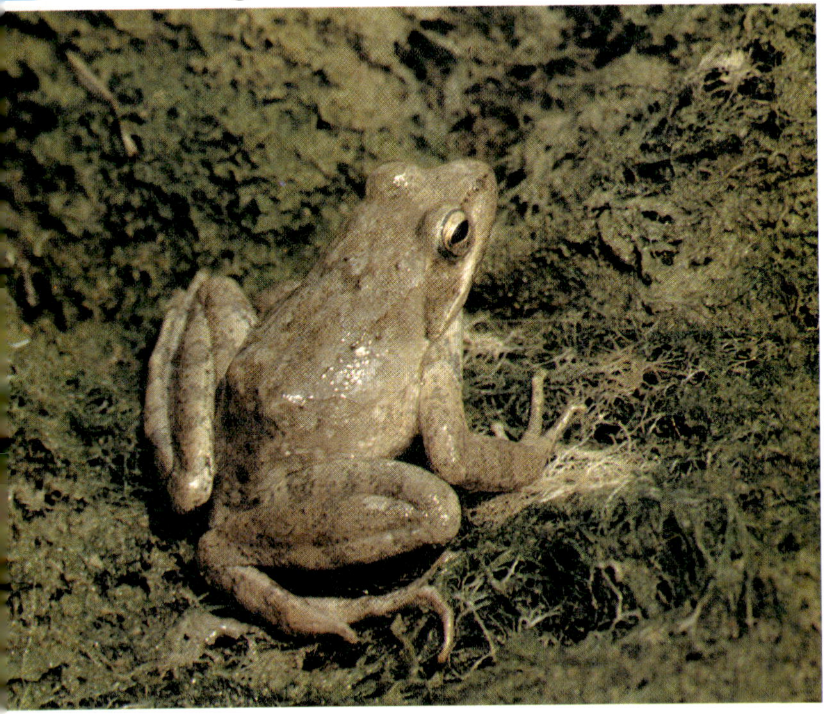

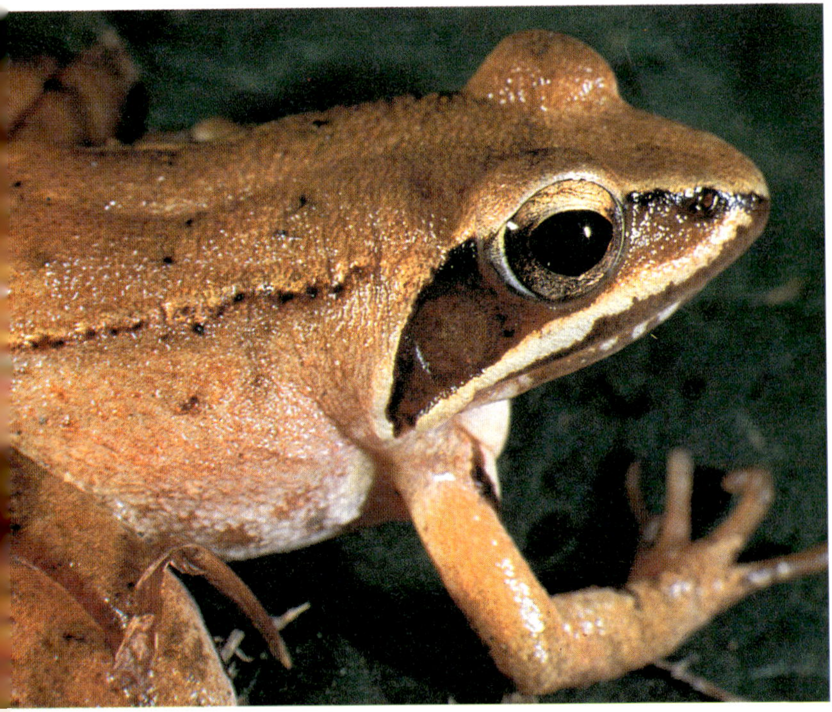

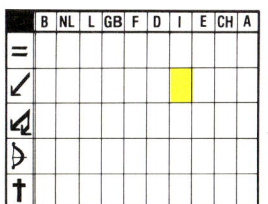

Stream Frog

	B	NL	L	GB	F	D	I	E	CH	A
=										
↙							▓			
◿										
◭										
†										

Very short call reminiscent of *Hyla arborea*. No distinct eardrum, no pinkish underparts. Distance between nostrils larger than distance between nostril and eye. Italian individuals smaller than Balkan animals.

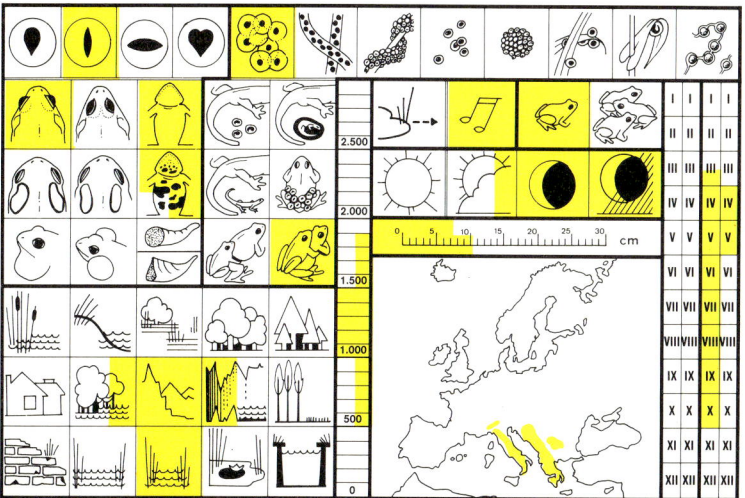

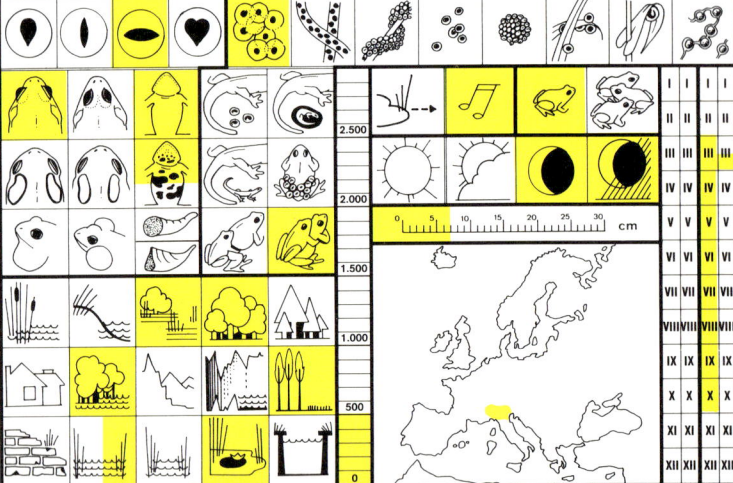

Italian Agile Frog

	B	NL	L	GB	F	D	I	E	CH	A
=										
↙										
◹										
◗										
✝										

Only in northern Italy. Underparts often with a pink flush. Distance between nostrils smaller than distance between nostril and eye. Lays 100–700 eggs.

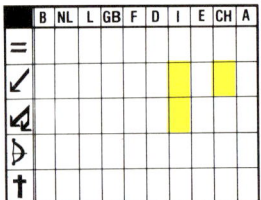

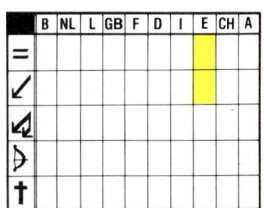

Iberian Frog

	B	NL	L	GB	F	D	I	E	CH	A
=										
↙										
◿										
▷										
†										

Hibernates and mates in the same water. Very fast and jumps well.

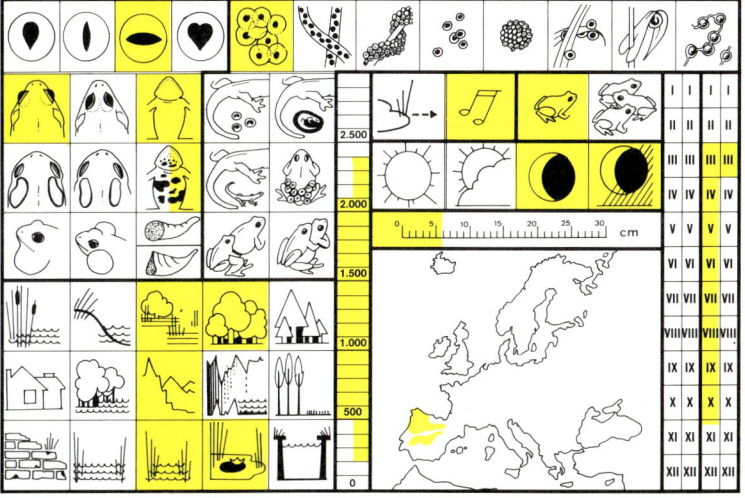

The Green Frog (*Rana esculenta*) Complex

Research over the past ten years indicates that there are several forms of the Green Frog. Basically there are two species: the Pool Frog (*Rana lessonae*) and the Marsh or Lake Frog (*R. ridibunda*). *R. perezi* was until recently considered a subspecies of *R. ridibunda*. According to most theories, *R. ridibunda* and *R. lessonae* came into contact after the ice ages, creating a hybrid: the Edible Frog (*R. esculenta*). However, *R. ridibunda* and *R. lessonae* seem to prefer different habitats, so are rarely seen together. *R. esculenta* occurs with either of them, but mostly *R. lessonae*. To breed successfully the Edible Frog must interbreed with one or other of the parent species, since *R. esculenta* x *R. esculenta* hybridisation produces unviable young. However, after hybridisation with *R. lessonae* or *R. ridibunda,* the offspring is always of the *R. esculenta* type. In some cases a pure population of *R. esculenta* may remain in existence.

Green frog

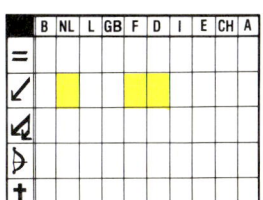

Marsh or Lake Frog

	B	NL	L	GB	F	D	I	E	CH	A
=										
⟋		■		■	■					
⟍										
⇗										
†										

Very shy species. Vocal sacs with grey pigments. Often attacks relatively large prey. Emits typical Green Frog croaks, but more rapid and higher-pitched than *R. lessonae* and *R. esculenta*. Up to 12,000 eggs. Introduced for frog's legs in many places.

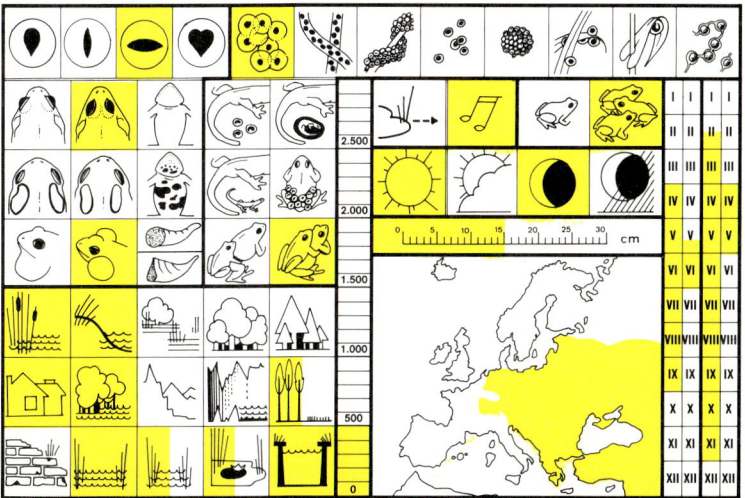

 Rana perezi

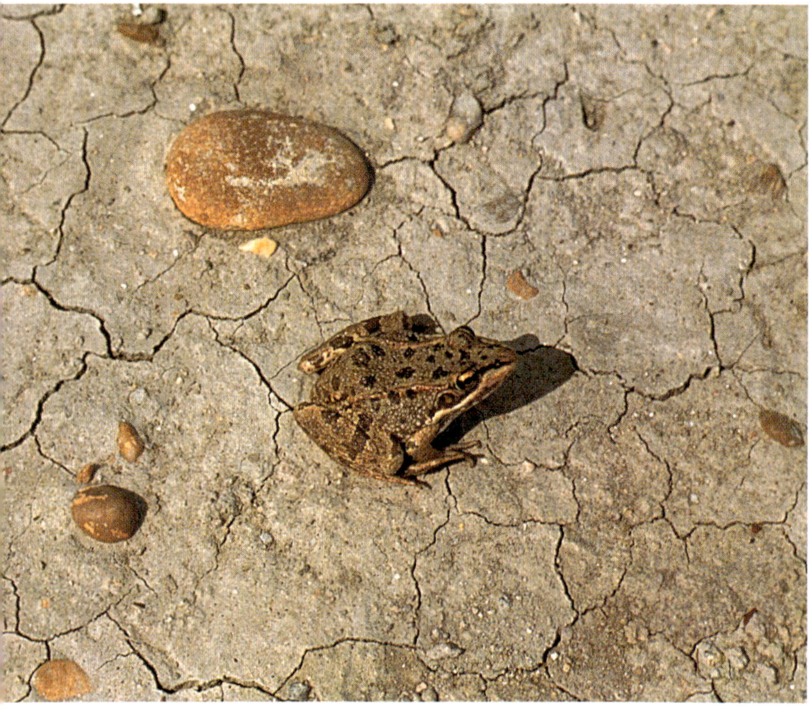

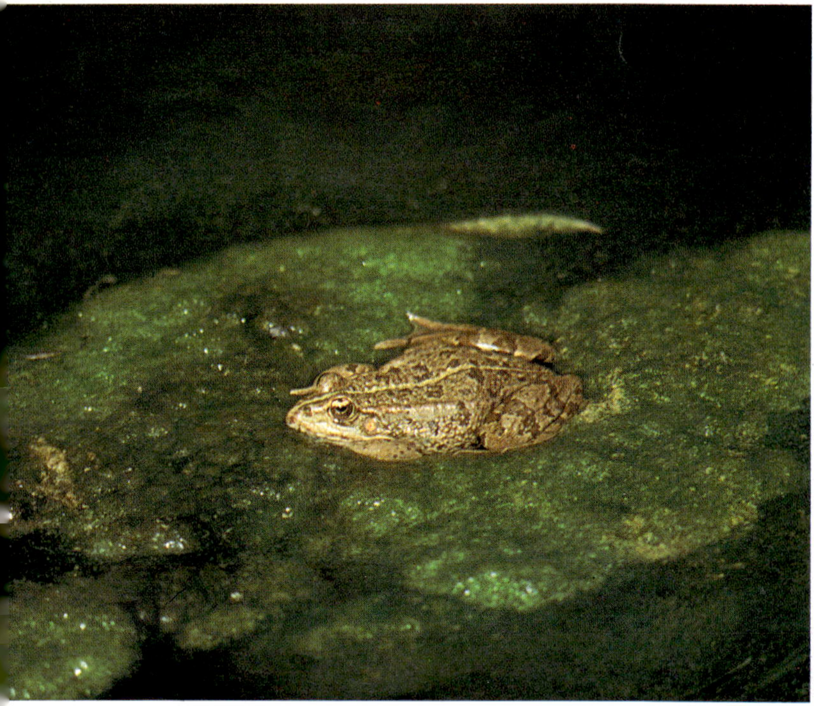

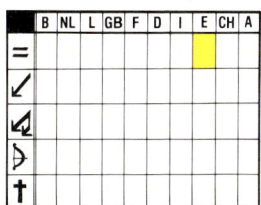

Southern Marsh Frog

	B	NL	L	GB	F	D	I	E	CH	A
=										
∕										
◁										
▷										
†										

Lays up to 10,000 eggs. For other characteristics see *R. ridibunda*.

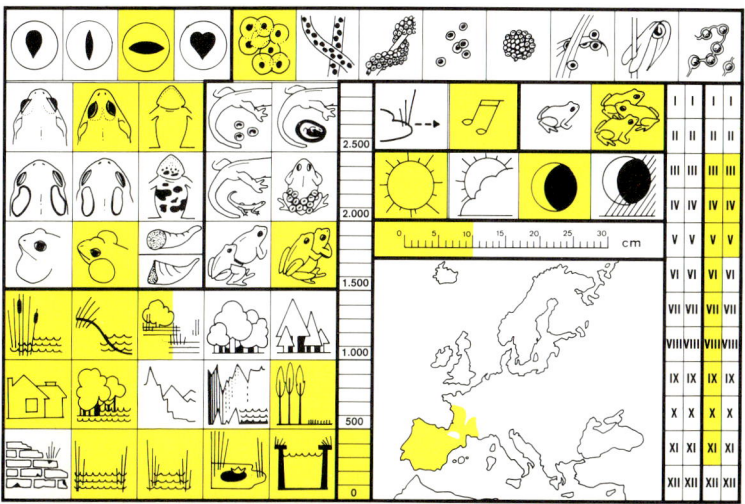

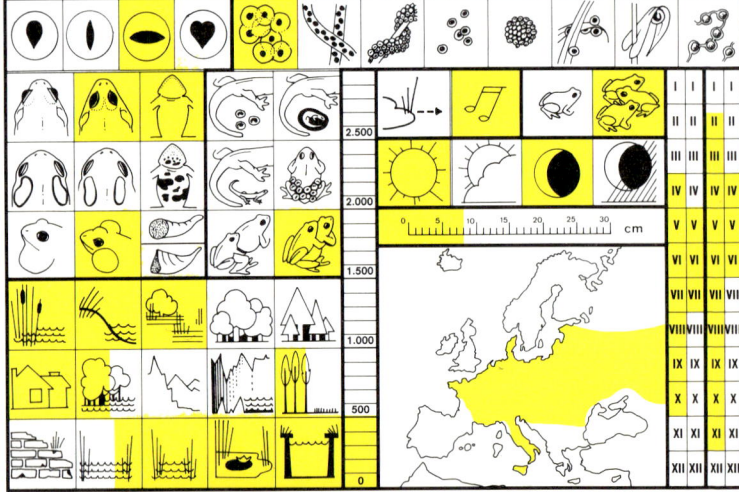

Pool Frog

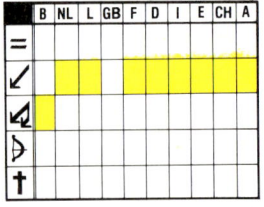

	B	NL	L	GB	F	D	I	E	CH	A
=										
↙										
◩										
♦										
✝										

Smallest Green Frog. Has white vocal sacs. Territory-forming less distinct than in other Green Frogs.

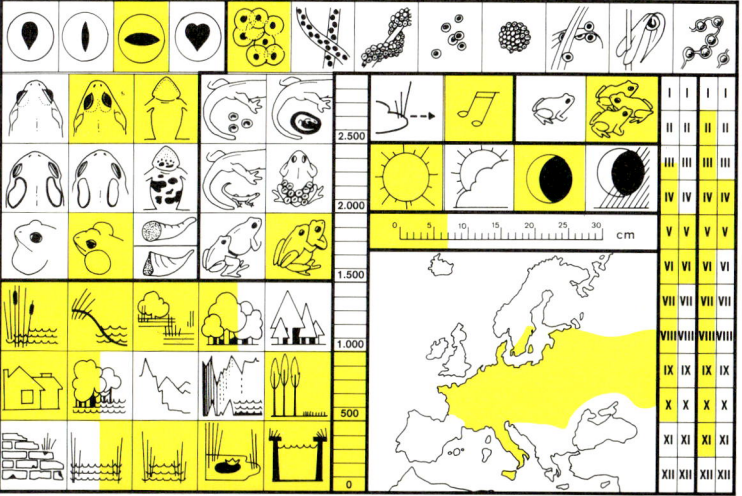

Edible Frog

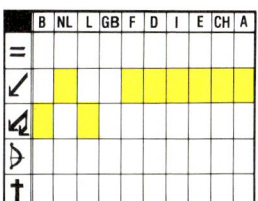

	B	NL	L	GB	F	D	I	E	CH	A
=										
↙										
↙										
⊅										
†										

This hybrid stands midway between *R. ridibunda* and *R. lessonae*. Stronger tendency to migrate than both other species. Croaks in chorus. Captured on a large scale for laboratories and frog's legs.

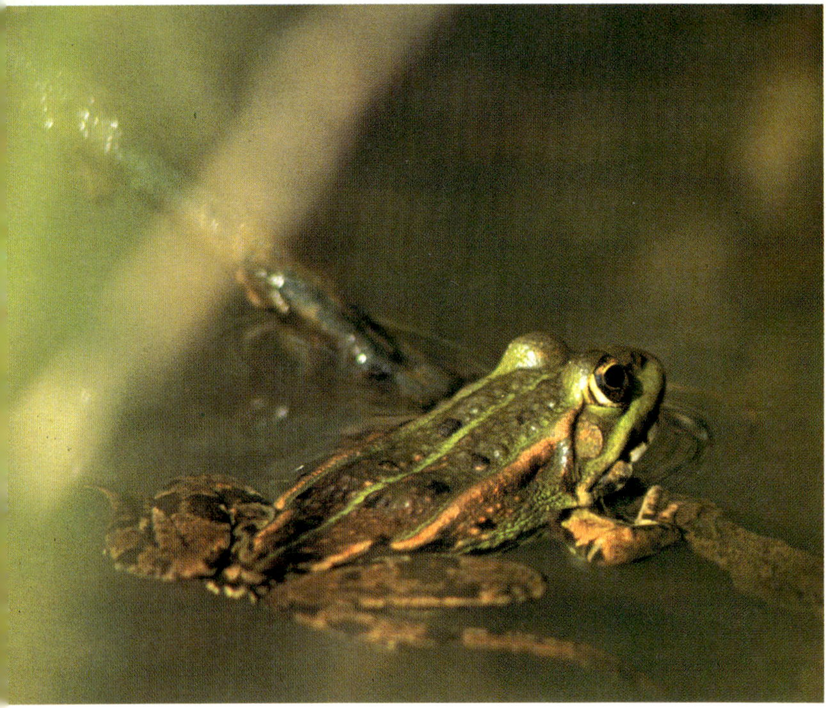

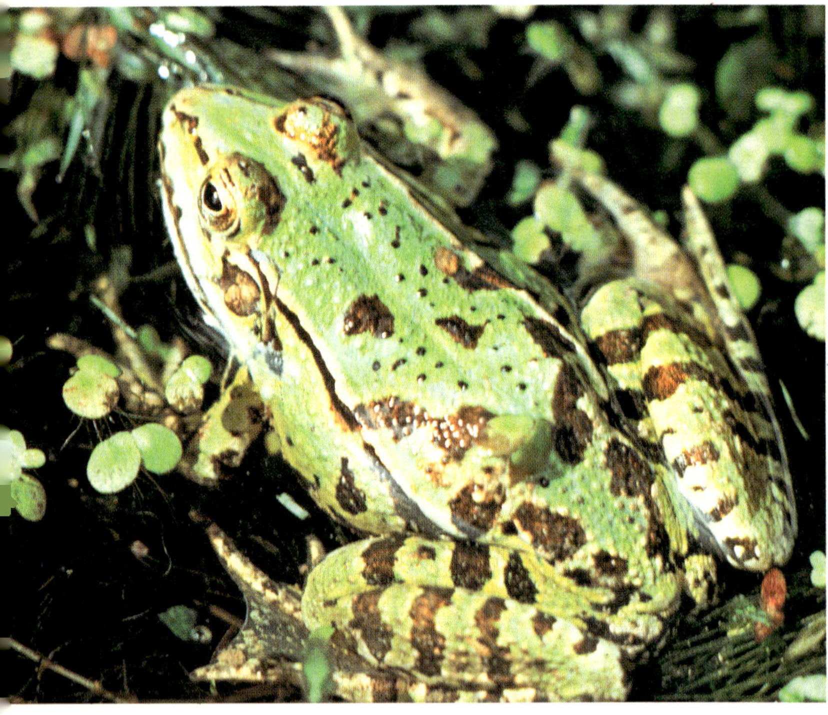

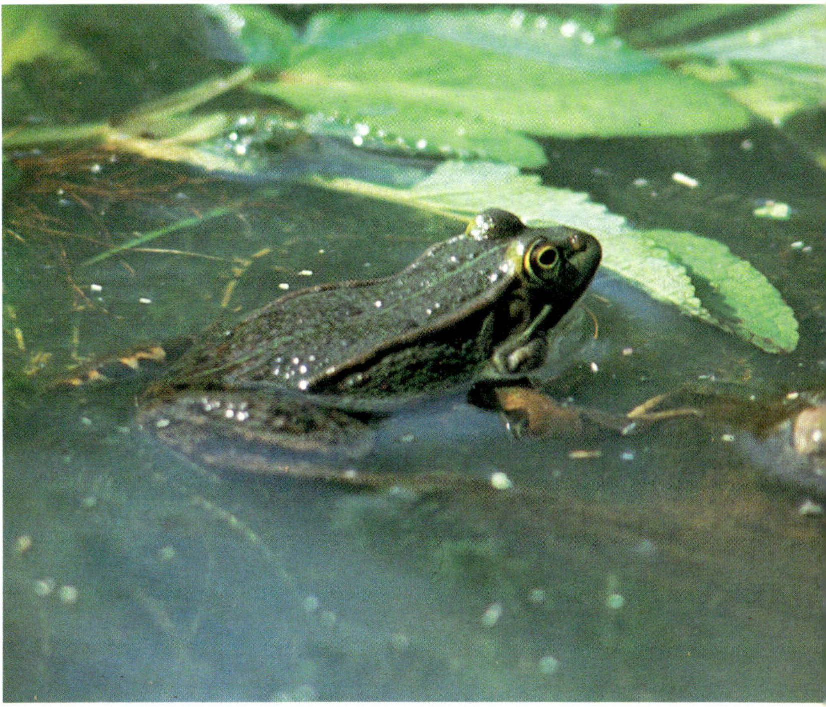

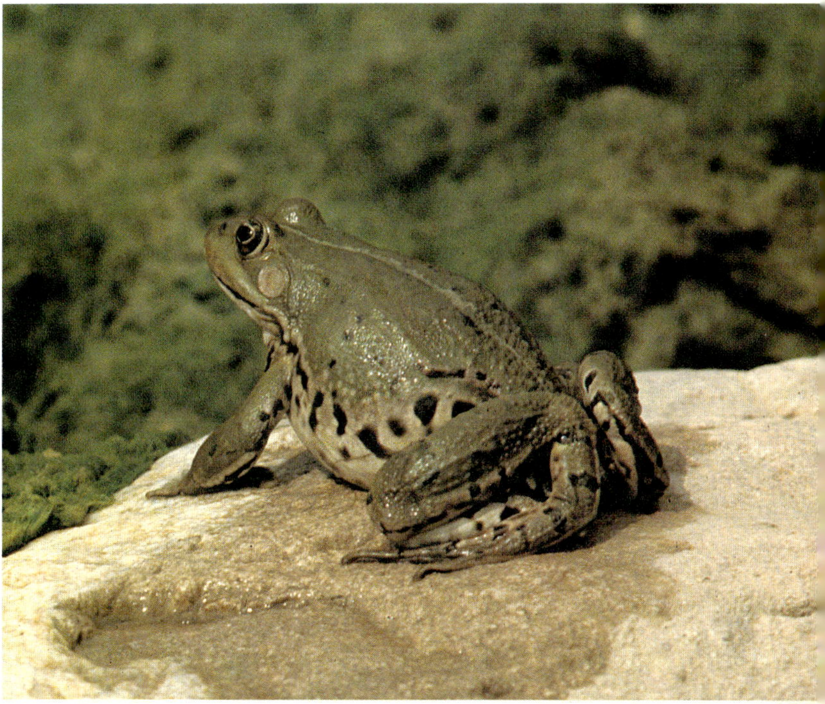

American Bull Frog

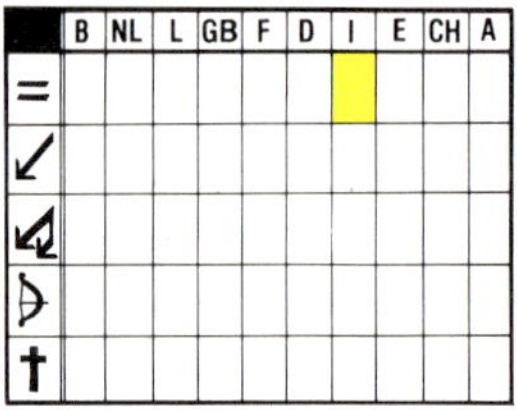

	B	NL	L	GB	F	D	I	E	CH	A
=										
↙										
◁										
▷										
†										

Introduced into northern Italy (Po Valley) during the forties. Locally threatens native amphibians. Voice: a deep groaning as of a cow.

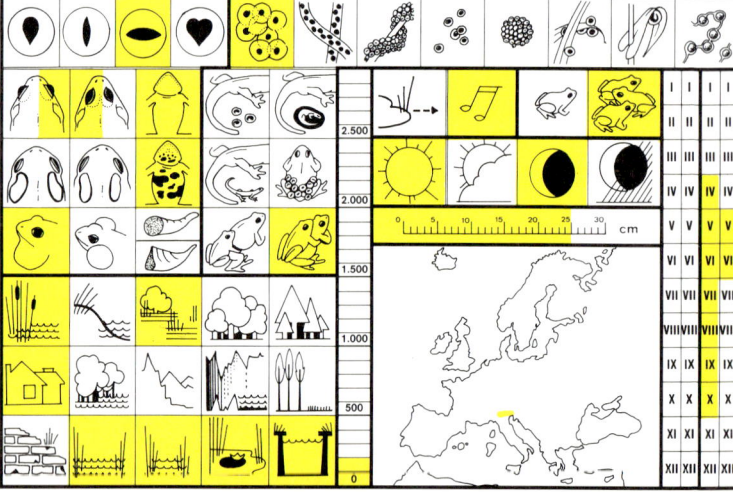

PHOTOGRAPHING AMPHIBIANS

A FEW PRACTICAL TIPS

The best time to make a photographic record of amphibians is the mating season which is mostly in the spring. Then you can be sure to find amphibians in or near water for egg-laying. What's more, the animals are very numerous then and not as shy as usual. They have their nuptial dress, especially conspicuous in newts, and you can get good pictures of certain species inflating their vocal sacs (*Hyla, Bufo viridis, B. calamita*).

Photographic Equipment

The best camera to use is a single-lens reflex (picture size 24 x 36mm), fitted with a small telephoto or a high-quality macro zoom to photograph the animals from a distance without disturbing them. A bellows or a set of extension rings with a 135mm or 300mm lens is good for close-up work.

For sharpest pictures use shutter speeds of approximately 1/125 or 1/250 and apertures preferably not above F/5.6 or F/8.

A tripod is recommended to avoid vibration. It is the only practical way to photograph motionless animals at shutter speeds slower than 1/125.

Always use flash if possible — even in daylight (at night you don't have a choice). Flash freezes the picture and makes it possible to shoot without using a tripod and with greater depth of field. It also gives you just enough light of equal colour temperature (5,500–6,000K). Amphibians are not generally frightened by flashes — certainly less so than birds or mammals.

For best results use two flashes or a ring light. The only drawback is that you have to take all these accessories along, though you should also be careful of short-circuits. This is not an imaginary risk in the damp environments amphibians prefer. To be safe, seal up the contacts with insulating tape and wrap the camera in plastic foil.

Films

In principle all types of film can be used and your choice is often a matter of personal preference. For flash pictures a 50 ASA film will give outstanding results, whereas daylight pictures require a 100 or 200 ASA film. For colour illustrations in books or magazines colour slides are essential.

The Photographer

Before starting, you should get to know the species you want to photograph as well as possible. You should wear wellingtons or thigh boots and carry some protection against rain. Try to disturb the animals as little as possible and avoid causing damage to banks and vegetation. If you have to catch hold of an animal, do so with a wet hand and no longer than needed. Eggs, tadpoles and newts can best be watched in a jamjar, a plastic bag or a small portable photo-aquarium.

Equipment Used for This Guide

Most pictures by the author and the other photographers were taken with Minolta reflex cameras (XD7, XG2, SRT 101, XM), equipped with a 135mm or 300mm telephoto fixed on bellows or extension rings. Salamanders and some toads were photographed using lenses of shorter focal lengths.

All prints were made from colour slides (Agfa CT18 and CT19, and Kodak Ektachrome and Ektachrome Professional).

The author at work

THE PROTECTION OF OUR AMPHIBIANS

The section on endangered species (pages 27–32) shows the serious-ness of threats to amphibians in Europe and the urgent need for drastic change, if we do not want a number of species to die out quickly.

International Conventions and National Legislation

The Washington Convention of 1973 (FITES)

This convention on trade in endangered species includes three categories of animal and plant species. The first list includes those species that are threatened with irreversible extinction and that must not be dealt in. The second list comprises species that are more or less endangered; trade in these species has been restricted by a system of licences. Finally there is a list of animals that are not endangered in their entirety but only in some countries. Dealing in these species is for-bidden when they come from these countries. European amphibians do not figure on any of these lists, though they should certainly be placed on the third.

The Bern Convention of 1979

This convention on the preservation of wild plants and animals and their natural environment was signed by 20 European countries and the EEC on 19 September 1979. On 7 May 1982 it was ratified by the EEC countries (9th ratifying party) and on 1 September it took effect for all ten EEC countries. This convention provides for strict protection of 400 endangered animal species, the ratifying countries committing themselves to take the necessary measures in order to protect the habitats in which these species live. This is of the utmost importance if we want to save our amphibian fauna from extinction.

The following amphibian species are strictly protected by the Bern Convention:

Appendix II: Animals — Amphibia

Salamanders and Newts: *Salamandra (Mertensiella) luschani*
(Luschan's Salamander)
Salamandrina terdigitata (Spectacled
Salamander)
Chioglossa lusitanica (Golden-striped
Salamander)
Triturus cristatus (Warty Newt)
Proteus anguinus (Olm)

Frogs and Toads: *Bombina variegata* (Yellow-bellied Toad)
Bombina bombina (Fire-bellied Toad)
Alytes obstetricans (Midwife Toad)
Alytes cisternasii (Iberian Midwife Toad)
Pelobates fuscus (Common Spadefoot)
Pelobates cultripes (Western Spadefoot)
Bufo calamita (Natterjack)
Bufo viridis (Green Toad)
Hyla arborea (Common Tree Frog)
Rana arvalis (Moor Frog)
Rana dalmatina (Agile Frog)
Rana latastei (Italian Agile Frog)

Also protected by the same convention are all other species not mentioned (Appendix III).

National legislation

Apart from these international conventions some countries have national legislation that protects amphibian fauna — though usually to a limited extent.

Austria: all amphibian species legally protected.
Belgium: the royal decree of 13 June 1973 limits the capture of frogs to private 'cultivated' ponds, the exploiter of which must have a licence. Unfortunately there are several exceptions: capture for export, scientific research, and local interest. The species concerned are not specified.
Belgium (Flanders): the royal decree of 22 September 1980 protects all species. The catching or disturbing of all species is strictly forbidden. Nor must the environment of the animals be disturbed (with the exception of catching Green and Common Frogs in private 'cultivated' ponds).
Belgium (Wallonia): the royal decree of 30 March 1983 protects all species (as in Flanders). Yet section 7 allows the capture of *Rana esculenta* and *Rana temporaria* (under the same conditions as above).
Czechoslovakia: a great number of species are legally protected.
France: in this country amphibians are protected by the act of 24 April 1979.
German Democratic Republic: with the exception of *Rana temporaria, Rana esculenta, Rana arvalis* and *Rana ridibunda* all species are protected. With licence, limited capture is permitted in order to keep the animals in captivity, but not for consumption.
Great Britain: British herpetofauna are strictly protected. Legally the protection is laid down in section 69 of the Wildlife and Countryside Act of 1981.
Greece: no specific legislation. Export is possible through a system of licences.
Hungary: some 30 amphibians and reptiles are protected, with the exception of frogs in private ponds (with a licence).
Ireland: all species protected by the Act of 1 November 1980.
Italy: except for the obligations arising from the Bern Convention, there is no national legislation to protect amphibians. Even the Act on

fauna protection and the hunting regulation (27 December 1977) does not mention them. Moreover only a few of the 19 regions have protective measures. These provincial acts (Trento, 25 July 1973; Bolzano, 13 August 1973; and the Venice Region, 15 November 1974) contain prohibition orders covering the taking of eggs, larvae and adults and in some cases they limit the capture of frogs.

Luxembourg: amphibians are protected by the Act of 29 July 1965, supplemented by the regulation of 3 November 1972. The legislation also attempts to protect habitats.

The Netherlands: the Nature Protection Act of 6 August 1973 protects all native amphibians and reptiles. Apart from the prohibition on catching, killing or possessing amphibians, there is a prohibition on disturbing animals or their nests. Unfortunately no direct habitat protection is afforded.

Poland: a number of species are protected.

Portugal: in principle all species are protected.

Rumania: no data.

Spain: most amphibian species are protected by the Act of 3 March 1981.

Switzerland: in general amphibians are protected by the federal Act of 1 July 1966 and in particular by the supplementing decree of 27 December 1966, which also pays attention to habitat protection and to countering habitat pollution.

West Germany: here the states (Länder) have their respective regulations, but this will change according to the Bern Convention. In the country as a whole the Acts of 18 March 1936 and 16 March 1940 have been supplemented by the federal decree of 1 September 1980, which protects all amphibian fauna. Special regulations are in force in Bavaria (Act of 29 June 1962), Baden-Wurtemberg (6 June 1963) and Hessen (8 March 1968, 10 July 1968 and 5 October 1970).

USSR: according to a Russian 'red data list' a number of species are legally protected.

Yugoslavia: no specific legal protection. In the state of Slovenia the Olm (*Proteus anguinus*) is legally protected.

Practical Measures

It goes without saying that this incomplete and often non-existent legislation offers insufficient protection for amphibians. In countries with the toughest legislation amphibians are still declining sharply. Laws prove difficult to apply and often have little effect.

It does not make sense to prohibit the capture or disturbance of rare species if their habitats can be destroyed at whim. Moreover many people are still ignorant of the problems facing amphibians, even in this age of greater awareness of the need for nature conservation.

There are only two possible ways to save what amphibians we still have. Firstly by establishing sanctuaries (in the fullest sense of the word) and secondly by informing the public at large. The RANA Group, which conceived this guide, was founded to achieve this dual objective (see page 123).

Establishing sanctuaries

Existing habitats with amphibian populations must be protected, if possible by creating a legally recognised sanctuary, even if it is a very small one. In the Netherlands some of the smallest sanctuaries in the world have been created for amphibians (*Alytes obstetricans* and *Bombina variegata*, Midwife Toad and Yellow-bellied Toad). In other countries too, especially Britain, West Germany and Switzerland, serious efforts have been made in the same direction. For rare species this is an urgent matter. The ideal solution would be the creation of buffer-zones around such sanctuaries and the connections between them.

Special attention should be devoted to the quantitative and qualitative preservation of as great a number of wetlands as possible. Since every amphibian species has its own specific requirements, the preservation of a wide variety of ponds and other habitats is of the utmost importance. These steps should be taken for all species — even those that are still quite common at the moment.

Within existing sanctuaries it is important that management serves the overall natural interest. Often it is focused on the protection of plants or birds, and in some cases this may have a negative impact on the amphibian fauna. Measures in favour of mammals or birds are not necessarily favourable to salamanders, newts, frogs or toads.

If it is essential to preserve a particular species in a sanctuary, the measures required could cause conflicts with ornithological or botanical interests. Mostly it is a matter of priorities, and in the restricted sanctuaries compromises are usually impossible. In such cases a population may be saved from extinction by creating artificial habitats, such as raising small slopes for Midwife Toads or digging new spawning ponds if the old ones are polluted or have become separated from the hibernation habitats. In Switzerland excellent results have been achieved in this way.

To preserve good spawning places it is absolutely essential that legislation protects the habitat. Legal requirements should preserve the physical and chemical qualities of the water — both surface water and subsoil water, including caves, springs and temporary pools.

The basis for all of this should be a thorough and complete census of amphibian populations and suitable breeding ponds. Such surveys have been carried out in Switzerland and West Germany for quite some time (Feldman), but in other European countries this is not taking place on a sufficiently large scale.

In many cases it may be important to make a list of amphibians that used to occur in the recent past, for which purpose the two best techniques are consulting collections in museums and asking older people, especially farmers. It is best to use photographs, but treat the answers with caution since numbers, size and species are often exaggerated.

On the basis of this data, area management can be properly undertaken. In this connection ecological research, by species and by population, is also of great importance. Specific measures can be taken, such as the fencing-in or temporary closing of roads that are crossed by migrating amphibians. However long-term solutions — such as the construction of tunnels or gratings — are better.

Occasionally, on the basis of such data, it may be possible to re-introduce individuals into empty, isolated habitats, when whatever caused the original disappearance has stopped and when animals from the surrounding neighbourhood are used. This is actually a very controversial issue among researchers. Only under the strictest conditions and after thorough study can it be justified in certain instances. Mammal experts are greater advocates of this idea than amphibian experts, possibly because so many amphibian re-introductions have ended in total failure.

Informing the public at large

It is of the utmost importance to make the public aware of amphibians on a wider scale than at present. This means providing a greater insight into what an amphibian is, how and where it lives, and the part it plays in the different ecosystems. This in itself will dispel many superstitions and negative attitudes, and make way for the same sort of sympathetic interest that exists for birds and mammals. Simply helping migrating amphibians cross roads is important from an educational point of view — and the schools which have engaged in such work have done so with great enthusiasm. Schools have an important role to play and a lot can be achieved in co-operation with nature conservation societies.

Mass media such as television, radio, newspapers and magazines are excellent ways to educate the public. In addition it is very important for nature conservation societies to pay proper attention to amphibians, both in the management of their sanctuaries and by promoting amphibians in the sanctuaries themselves, with leaflets, nature rambles, and personal information to visitors who may inadvertently cause damage to amphibian populations.

What YOU can do to protect amphibians

Here's how you can help. In the first place, of course, you should observe the nature protection laws. But you can go much further and actively participate in collecting data on amphibians and their breeding places.

Farmers should be made aware of the presence of amphibians on their land. Children can learn about the problems of salamanders, newts, frogs and toads at school, especially with practical class assignments on the subject. Local environmental groups can assist in organising actions such as taking amphibians across the road in spring. Local authorities should be made aware of the presence and the problems of amphibian populations in their districts.

If you live in or near an area where natural pools and ponds are disappearing, you can dig substitute ponds in your garden — without fish of course. However, you must never catch animals elsewhere and release them in your garden pond. Adults will simply return to the place from which they were taken.

Below is a diagram of a suitable pond (Fig 4). It can be ready-made out of plastic, plastic foil or polyester. Concrete is less suitable and is also more expensive and not as simple to construct. Consult any of the many publications on the subject for tips on construction. The deepest

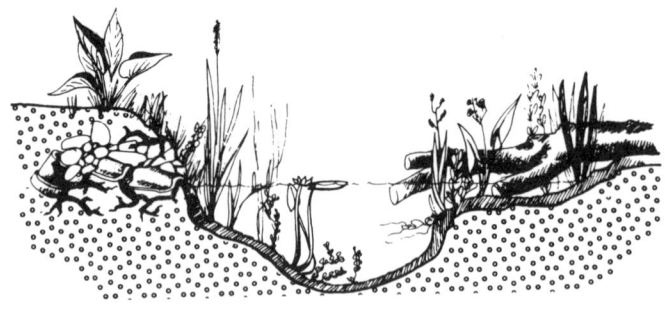

point should be about 50cm and the slopes should not be steep but gradual. The pond should have some aquatic vegetation and, if *Bufo bufo* (Common Toad) is present, reeds or other upright vegetation to enable it to wind the egg strings around them. The situation should be quiet and sunny. And then you simply wait and hope they will come . . .

If need be you can also create a hibernation place. A layer of mud at the bottom of the pond will benefit the species that hibernate underwater (*Rana ridibunda, R. esculenta, R. temporaria* and others). A compost heap, loose leaves, branches, loosely piled logs or a quiet place with a loose soil structure may be useful to species that hibernate on land.

LIST OF ADDRESSES OF ORGANISATIONS ENGAGED IN THE STUDY AND PROTECTION OF AMPHIBIANS

The RANA group was founded in order to provide greater study, surveys and actual protection of amphibians and reptiles — and keep conservationists and the public at large well-informed.

We try to bring separate studies together and provide better distribution of the results in cooperation with larger organisations and institutions for the study and protection of nature. At the moment West European attention is being focused on activities and campaigns in those countries where amphibian and reptile protection is virtually non-existent: Italy and Belgium. In countries where other groups are engaged in similar activities, we try to cooperate to as great an extent as possible (see list of addresses).

Anyone interested in these activities or prepared to engage in survey work, campaigns to help migrating amphibians cross roads, and so on, should contact the nearest RANA branch listed below.

The fact that you are reading this shows you are interested in this topic.

Perhaps, fascinated by the wonderful world of amphibians and aware of their decline, you would like to do something about the situation. If so, all you have to do is contact RANA.

Austria
Societas Europeae Herpetologica
Naturhistorisches Museum
Postfach 417
A – 1014 WIEN

Belgium
RANA – Belgium
Populierenlaan 17
B-1980 TERVUREN

or
Eglantierlaan 39
B-2630 AARTSELAAR

Czechoslovakia
Herpetological Section – Czechoslovak Zoological Society
Department Zoologie
Fakulteit Wetenschappen
Palacky University
Leninova 26
77 146 OLOUMOUC

France
RANA – France
Les Vignes du Mas n°16
F – 16150 CHABANAIS (F)

Société Batrachologique de France
Laboratoire des Reptiles et Amphibiens
Muséum National d'Histoire Naturelle
25 Rue Cuvier
F – 75005 PARIS

Société Herpétologique de France
Université de Paris VII
Laboratoire d'anatomie Comparée
2 Place Jussieu
F – 75230 PARIS Cédex 05

German Democratic Republic
Kulturbund der DDR – Abteilung Natur und Umwelt
DDR – 1030 BERLIN
PSF 34

Great Britain
Association for the Study of Amphibia and Reptilia
The ASRA Rooms
c/o Cotswold Wildlife Park
BURFORD
Oxon OX8 4JW

British Herpetological Society
c/o Zoological Society of London
Regent's Park
LONDON NW1 4RY

Conservation Committee of Societas Europaea Herpetologica
136 Estcourt Road
Woodside
LONDON SE25 4SA

International Herpetological Society
27 St Thomas Close
Dartmouth Avenue
WALSALL
West Midlands WS3 1SZ

South-Western Herpetological Society
59 St Marychurch Road
TORQUAY
Devon TQ1 3HG

Thames and Chiltern Herpetological Group
The Youth Club
Narcot Lane
CHALFONT ST GILES
Buckinghamshire

Hungary
First Herpetological Conference of the Socialist Countries
Dr O. Gy. Dely
Zoological Department
Hungarian Natural History Museum
Baros u. 13
H – 1088 BUDAPEST

Herpetological and Terraristic Faculty of the Society of Natural Sciences
Tit Természettid Studio
Bocskai ut 37
H – 1113 BUDAPEST

Italy
RANA – Italy
Via Le Forchie
Casa Postale 41
I – 70017 PUTIGNANO (Bari)

The Netherlands
Werkgroep Amfibieën Nederland (WARN)
Kemperbergerweg 67
NL – 6818 RM ARNHEM

Lacerta
Commissie Bedreigde Amfibieën en Reptielen
Jagersdreef 144
NL-3972 XH DRIEBERGEN

Lacerta-Werkgroep Limburg
Meutestraat 7B
NL-6219 BH MAASTRICHT

Herpetogeografische Dienst
W. Bergmans
Pieter Pauwstraat 10'
NL-1017 ZJ AMSTERDAM

Societas Europaea Herpetologica
Rijksmuseum voor Natuurlijke Historie
Postbus 9517
NL-2300 RA LEIDEN

Poland
Herpetological Section – Polish Zoological Society
Plac Weilkopolski 2/55
PL-61 – 746 POZNÀN

Spain
Associacion Iberica de Herpetofauna
Avenida de Logrono 23
E-MADRID 22

Grupo ed Estudid de los Anfibios y Reptiles Ibéricos
Centro Pirenaico de Biologia experimental
Apartado 64
E – JACA (HUESCA)

Switzerland
Koordinationstelle für Amphibien und Reptilienschutz in der Schweiz
Naturhistorisches Museum
Bernastrasse 15
CH – 3005 BERN

West Germany
Societas Europaea Herpetologica
Institut für Zoologie
Johannes Gutenberg Universität
D – 6500 MAINZ

Deutsche Gesellschaft für Herpetologie und Terrarienkunde
Natur Museum Senckenberg
Senckenberganlage 25
D – 6000 FRANKFURT I

USSR
Union of Herpetologists of the Socialist Countries
Zoologicheskii Institut
Akademiya Nauk SSSR
LENINGRAD V – 164

BIBLIOGRAPHY

Angel, F., *Faune de France: Reptiles et Amphibiens*, Lechevalier, Paris (1946)

Arnold, E.N., J.A. Burton & D.W. Ovenden *A Field Guide to the Reptiles and Amphibians of Britain and Europe* Collins, London (1978)

Andrada, J., *Guía de Campo de los Anfibios y Reptiles de la Península Ibérica* Omega, Barcelona (1980)

Ballasina, D., Ça va mal avec nos grenouilles, crapauds et salamandres! I: Animalia *Bull. Trim. Soc. Protection Anim.* Veeweyde, Bruxelles (1982)

Beebee, T.J.C., Observations Concerning the Decline of the British Amphibia *Biol. Conserv.*, 5: 20–24 (1973)
idem 1975 Surveys of British Amphibians and their Habitats *Environmental Conservation*, 2: I 36

Bergamans, W. & A. Zuiderwijk, Amfibieën en Reptielen in Nederland Wet. Mededelingen *Kon. Ned. Natuurhist. Ver.*, 139 (1980)

Brockleman, W.Y., *Ecology*, 50(4): 632–644 (1969)

Bruno, S., Anfibi e Rettili di Sicilia Catania, *Atti Acad. Sc. Nat.*, Ser. VII II: I–144 (1970)
idem 1973 Problemi di Conservazione nel Campo dell'Erpetologia Bari, *Atti II Simp. Naz. Conserv. della Natura*, II: 117–226

Bruno S., E. Burattini, A. Casale, Il Rospo Bruno del Cornalia Bari, *Atti IV Simp. Naz. Conserv. della Natura*, II: 33–56 (1974)

Capula, M., Prima che Gli Anfibi Scompaiono *Panda*, Pubbl. mens. WWF Italy, Roma, XVI I: 3–5 (1982)

De Fonseca, Ph., Inheemse Amphibia en Reptilia *COBRA* 21/23: 49–58 idem 1979 Menacés, Mal Aimés, Méconnus, nos Amphibiens et Reptiles *Réserves Naturelles* no 2: 7–14 (1978)

Dottrens, E., *Amphibiens et Reptiles. Merveilles de la Nature* Delachaux et Nieslé, Neuchâtel (1963)

Feldmann, R., Ergebnisse vierzehnjähriger quantitativer Bestandkontrollen an *Triturus* Laichplatzen in Westfalen *Salamandra*, 14,3: 126–146 (1978)

Fretey, J., *Guide des Reptiles et Batraciens de France* Hatier, Paris (1975)

Gelder, J.J. van, Uitzetten van dieren — zinnig of onzinnig? *De Levende Natuur*, 80, 5: 105–111 (1977)

Honegger, R.E., *Threatened Amphibians and Reptiles in Europe* Nature and Environment series No 15, Council of Europe, Strasbourg (1980)

Kowalewski, L., Observations on the Phenology and Ecology of Amphibia in the Region of Czestochowa *Acta, Zool. Cravoc.*, 19: 391–438 (1974)

Mertens, R., *Kriechtiere und Lurche* Kosmos, Stuttgart (1975)

Parent, G.H., *Atlas Provisoire Commenté de l'Herpétofaune de la Belgique et*

du Grande Duché du Luxembourg Naturalistes Belges, Bruxelles (1979)
idem *Protégeons nos Batraciens et Reptiles. Animaux menacés en Wallonie*
Région wallonne Jambes et Ed. Duculot Gembloux-Paris
pp1–172 (1983)

Petretti, F., *Animali in Pericolo di Estinzione* Musameci, Aosta (1980)

Pozzi, A., La Rana de lataste *SOS Fauna, animali in pericolo in Italia*
349–356 WWF, Roma (1976)

Riggio, S., Il Discoglosso in Sicilia *SOS Fauna, animali in pericolo in Italia*
417–464 WWF, Roma (1976)

Salvador, A., *Guia de los Anfibios y Reptiles Españoles* Icona, Madrid
(1974)

Sparreboom, M., *De Amfibieën en Reptielen van Nederland, België en
Luxemburg* Balkema, Rotterdam (1981)

Tortonese, E. & B. Lanza, *Piccola Fauna Italiana: Pesci, Anfibi e Rettili*
Aldo Martello Editore, Milan (1968)

Vermehren, K., *Kikkers en Padden* Balkema, Rotterdam (1977)

Wijnands, H.E.J., Distribution and habitat of *Rana esculenta* complex
in the Netherlands *Neth. J. Zool.*, 27: 277–286 (1977)

Witte, G.F. de, *Amphibiens et Reptiles de la Belgique* Mus. Hist. Nat.,
Bruxelles (1948)

ABOUT THE AUTHOR

Donato Ballasina is a 26 year-old biologist with a special interest in herpetology. His research work has taken him to most European countries from the Netherlands to Italy, Yugoslavia and Greece — and further afield to the USA and Mexico.

He is well-known as the author of more than 100 articles, and works regularly with the Committee for the Protection of Birds and the World Wildlife Fund.

He is also a founder of the RANA GROUP, created to protect and promote knowledge of 'Reptiles and Amphibians in Nature' throughout Europe.

ACKNOWLEDGEMENTS

The author would like to thank his parents, for their understanding and their ever-present encouragement, Philippe De Fonseca, to whom thanks for his support and his stimulation have unfortunately to be a posthumous tribute.

Also all nature-lovers and all members and contributors to the RANA Group, in the first place to his wife Veerle and friends Frederik De Wilde, Sandro Frisenda and Nico Willemsens for their help with this work and for their efforts in preserving our endangered herpetofauna for the future.

Many thanks to all friends, colleagues and those who have kindly contributed to this book in order to popularise knowledge and protect amphibians (alphabetically): Noemi Ballasina, Ibrahim Baran, Alain de Fonseca, Kurt Grossenbacher, Freddie Hordies, Stephanie Krebel, Michaël Lambert, Harry Mooijenkind, Henk Strijbosch, Jef van Ammel, Niels Van Esterik, André Van Hecke, Harrie Wijnands, Henriette Wolfs.

Thanks are also expressed to John A. Dudley for his help in revising the English translation of the text.

INDEX

References in *italic type* refer to drawings of larvae
References in **bold type** refer to colour plate numbers